ANSWER BOOK FOR WORKBOOK SECTION FOR

Geometric Dimensioning and Tolerancing Textbook and Workbook

(per ASME Y14.5-2018)

ANSWER BOOK FOR WORKBOOK SECTION FOR Geometric Dimensioning and Tolerancing Textbook and Workbook (per ASME Y14.5-2018)

James D. Meadows

Bibliography

ASME Y14.5-2018 Dimensioning and Tolerancing

ASME Y14.5-2009 Dimensioning and Tolerancing

ASME Y14.5M-1994 Dimensioning and Tolerancing

ASME Y14.5.1M-2018 Mathematical Definition of Dimensioning and Tolerancing Principles

ASME Y14.43-2011 Dimensioning and Tolerancing Principles for Gages and Fixtures

ASME Y14.2 Line Conventions and Lettering

ASME Y14.3 Orthographic and Pictorial Views

ASME Y14.41 Digital Product Definition Data Practices

ASME Y14.100 Engineering Drawing Practices

ASME Y14.8 Castings and Forgings

All of the Y14 standards above are published by The American Society of Mechanical Engineers, New York, New York.

Geometric Dimensioning and Tolerancing-Applications, Analysis, Gauging & Measurement [per ASME Y14.5-2018] by James D. Meadows co-published by The American Society of Mechanical Engineers, New York

Geometric Dimensioning and Tolerancing-Applications, Analysis & Measurement [per ASME Y14.5-2009] by James D. Meadows published by The American Society of Mechanical Engineers, New York

Workbook and Answerbook for Geometric Dimensioning and Tolerancing [per ASME Y14.5-2009] by James D. Meadows published in 2009 by James D. Meadows & Associates, Inc.

Tolerance Stack-Up Analysis for Plus and Minus Tolerancing and Geometric Dimensioning and Tolerancing by James D. Meadows published in 2010 by James D. Meadows & Associates, Inc.

Geometric Dimensioning and Tolerancing, Applications and Techniques for Use in Design, Manufacturing, and Inspection by James D. Meadows published in 1995 by Marcel Dekker, Inc., New York, New York.

Geometric Dimensioning and Tolerancing Workbook/Answerbook by James D. Meadows published in 1997 by Marcel Dekker, Inc., New York, New York.

Geometric Dimensioning and Tolerancing in 2007 by James D. Meadows published in 2007 by James D. Meadows & Associates, Inc., Hendersonville, TN.

Geometric Dimensioning and Tolerancing in 2007 Workbook and Answerbook by James D. Meadows published in 2007 by James D. Meadows & Associates, Inc., Hendersonville, TN.

Measurement of Geometric Tolerances in Manufacturing by James D. Meadows published in 1998 by Marcel Dekker, Inc., New York, New York.

Differences and Similarities between ASME and ISO Dimensioning and Tolerancing Standards by James D. Meadows published in 2006 by James D. Meadows & Associates, Inc.

DataMyte Handbook a practical guide to computerized data collection for Statistical Process Control by DataMyte Business Allen-Bradley Company, Inc.

Statistical Tolerancing, A Solution to Procurement Problems-US Army Tank Command.

ANSWER BOOK FOR WORKBOOK SECTION FOR GEOMETRIC DIMENSIONG AND TOLERANCING TEXTBOOK AND WORKBOOK (per ASME Y14.5-2018)

JAMES D. MEADOWS

James D. Meadows & Associates, Inc.

Hendersonville, Tennessee 37075

jamesmeadows@comcast.net

www.geotolmeadows.com

(615)924-3600

Published and distributed by:
James D. Meadows & Associates, Inc.
170 E. Main, D-137
Hendersonville, TN 37075
Phone: (615) 924-3600
JDMeadows@geotolmeadows.com
jamesmeadows@comcast.net
www.geotolmeadows.com

ISBN: 978-0-578-69252-4
Printed in the United States of America

1 2 3 4 5 6 7 8 9 10 Printing

Foreword

ABOUT THE BOOK:

This book is the Answer book for the Workbook Section of the Geometric Dimensioning and Tolerancing Textbook and Workbook (per ASME Y14.5-2018). It is the completion of that GD&T learning tool. It is meant for universities, companies and individuals. It provides answers to all of the worksheets to allow one to gage the reader's level of understanding.

It contains over 300 basic and advanced answers and illustrations to the problems in the workbook section of the textbook that can be worked through by individuals or teams.

This is an answer book that shows not only how to interpret and apply the concepts, but also conveys the logic critical to understand in order to design, dimension and tolerance products optimally. The sections covered include:

Section 1-Word Problems
Section 2-Basic Applications

Section 3-Interpretation and Analysis

Section 4-Tolerancing Parts & Assemblies

Section 5-Analysis and Mating Parts

Section 6-Wall Thickness & Composite Tol.

Section 7-New in Y14.5-2009 & 2018

Section 8-Advanced Applications

1. Terms and Definitions pertaining to Datums and Datum Features

1) Datums are
   b. simulated
   c. perfect
   e. derived from datum features
2) Features are
   a. real
   b. surfaces
   e. sometimes measurable for size
3) Datum features are
   b. real
   d. sometimes size features
   e. surfaces
4) Surfaces are
   a. features
   c. real
   e. flawed
   f. sometimes chosen to serve as datum features
5) Datum features should be chosen based on
   a. function
   c. repeatability
   d. accessibility
   e. representation of mating features
6) List five common physical datum feature simulators.
   surface plates, angle plates, chucking devices, V-Blocks and parallels
7) A dimension with a box around it like [100], is called?
   b. basic dimension
8) What is the [A] symbol called?
   a. datum feature symbol
9) What kind of tolerance is expressed by 50 ± 1?
   c. bilateral plus and minus tolerancing
10) What is the Ⓜ called when used after the geometric tolerance?
   d. maximum material condition symbol
11) What is third datum feature in the feature control frame called?
   b. tertiary
12) What is the Ⓜ called when used after a datum feature of size?
   d. maximum material boundary symbol
13) If a Ø2 appears in a feature control frame, it is
   d. the geometric tolerance
14) List at least two components (other than the perpendicularity symbol) that you feel are needed in a perpendicularity control. **a geometric tolerance and a datum reference**
15) A clearance fit implies the mating features, when produced within size limits,
   c. will never encounter material interference.
16) An interference fit implies the mating features, when produced within size limits,
   a. will always encounter material interference.
17) In a fixed fastener assembly condition, the geometric tolerance to be divided between the two parts containing the mating features is equal to
   b. the difference between the Maximum Material Conditions of mating features.

2. MMC and LMC

Please identify the Ⓜ and Ⓛ of these.

HOLES:

1)	62.5 ± 0.1	Ⓜ	**62.4**
		Ⓛ	**62.6**
2)	59-62	Ⓜ	**59**
		Ⓛ	**62**
3)	83.5 ± 0.05	Ⓜ	**83.45**
		Ⓛ	**83.55**
4)	69.5-70.5	Ⓜ	**69.5**
		Ⓛ	**70.5**
5)	1.245-1.265	Ⓜ	**1.245**
		Ⓛ	**1.265**

SHAFTS:

6)	7.5-8.1	Ⓜ	**8.1**
		Ⓛ	**7.5**
7)	71 ± 0.2	Ⓜ	**71.2**
		Ⓛ	**70.8**
8)	12.45-12.65	Ⓜ	**12.65**
		Ⓛ	**12.45**
9)	21.885-22.875	Ⓜ	**22.875**
		Ⓛ	**21.885**
10)	0.85-0.90	Ⓜ	**0.90**
		Ⓛ	**0.85**

3. Size Controls Form

1) Unless otherwise specified, individual features of size must
c. not violate an envelope of perfect form at MMC.

2) The concept discussed in Question 1 is often inspected with a
b. GO gage.

3) Least material condition on most parts is checked
a. at cross sections.

4) When checking for maximum material condition violations in a hole with size limits of Ø50 - 51, what size and type of gage would you use?
a. a full form plug gage with a diameter of 50

5) What effect on size measurements would adding the Independency symbol have? **It would change the envelope/GO gage requirement at MMC to a cross-sectional check requirement.**

6) Ø25 ± 0.1 AVG. This is a hole in metal. It is the I.D. of a thin walled tube.
What is the AVG. MMC? **Ø24.9** What is the AVG. LMC? **Ø25.1**

7) Ø.500 ± .005 This is a flèxible rubber shaft.
What is the MMC? **Ø.505** What is the LMC? **Ø.495**

8) .625 - .635 This is a slot milled into a thick rigid plate.
What is the MMC? **.625** What is the LMC? **.635**

9) .010 ± .002 This is a steel tab that is non-rigid.
What is the MMC? **.012** What is the LMC? **.008**

10) Ø100 ± 1
| ⊕ | Ø0Ⓛ | A | B | C |
This control refers to a glass shaft outside diameter. A hole runs through its center making it look like a drinking straw. It is mounted on a plate.
What is the MMC? **Ø101** What is the LMC? **Ø99**

How does the use of the LMC symbol change the way the size is measured? **An envelope of perfect form at LMC must be verified. No portion of the shaft may be inside of a perfect cylinder with a Ø99. The MMC of Ø101 is verified at cross-sections, instead of a GO gage.**

11) Ø37.5 ± 0.2
| — | Ø0.5Ⓜ |
This is a metal rod (shaft) that is twenty-five feet long.
What is the MMC? **Ø37.7** What is the LMC? **Ø37.3**

How does the use of straightness of the derived median line change the way size is measured? **All size measurements for both MMC and LMC are checked at cross-sections. The envelope of perfect form at MMC normally required and inspected with a GO gage is eliminated.**

12) 10.0 - 10.1 This forms two sides of a rectangular part. It is considered an external feature of size. It is rigid.
What is the MMC? **10.1** What is the LMC? **10**

4. Rules, Concepts, Characteristics and Untoleranced Dimensions

1) What purpose does referencing a datum in a feature control frame serve? **It stabilizes the part, eliminates various degrees of spatial freedom and controls either orientation or location or both orientation and location.**

2) Please categorize all the geometric characteristic symbols as:
INDIVIDUAL (not related to datums); RELATED (related to datums); INDIVIDUAL or RELATED:
Not related to datums: Straightness, Flatness, Circularity and Cylindricity.
Related to datums: Parallelism, Perpendicularity, Angularity, Circular Runout and Total Runout, Concentricity and Symmetry.
Either related or not related to datums: Profile of a Line, Profile of a Surface and Position.

3) When we check an individual feature for form, what are we comparing it to?
b. a perfect geometric form of itself

4) Once on a drawing, how can we tell if a feature is individual or related? **If related, a note or feature control frame will reference the relationship between the feature being constrained and the datum or datums to which it is related.**

5) Name three ways to eliminate the requirement for an envelope of perfect form at MMC for a feature of size. **1. With a note such as "PERFECT FORM AT MMC IS NOT REQUIRED." 2. Use of an average dimension such as 1200-1210 AVG. 3. Use of a least material condition symbol inside of the feature control frame for that feature. 4. Use of a straightness of the derived median line or derived median plane may be specified. 5. Use of the Independency symbol.**

6) If you do one of the three things stated in Question 5, can you exceed MMC at any cross sectional check? **Yes, with the use of AVG dimensioning, the size of the produced feature must only average within the size limits. A minimum of four measurements is taken at each cross section of the feature and then averaged.**

7) If you assign the LMC concept and symbology to the geometric tolerance in the feature control frame, where is the envelope of perfect form size requirement?
c. at LMC

8) Which material condition symbol or symbols (MMC, LMC or RFS implied) commonly provides the most geometric tolerance to draw from when used? **MMC and LMC symbols**

9) Which parts are considered to be most economical to manufacture?
a. those with features controlled at MMC

10) What material condition symbol provides for the easiest use of functional (receiver-type) gaging? **the MMC symbol**

11) Is it ever possible to use the Ⓜ symbol both for a feature of size and a datum feature of size in one feature control frame?
For example: | ⌖ | ⌀0.5Ⓜ | A | BⓂ | C | **Yes, this is quite common.**

12) When features are controlled RFS, they are given an original tolerance of form, orientation, profile, runout or location. Where can we draw extra geometric tolerance from?
c. we can't

13) As a minimum, the feature control frame must contain
b. a geometric characteristic symbol and a geometric tolerance.

14) Draw the symbols for the following geometric characteristic symbols and categorize them as form, orientation, profile, runout or position.

a. Flatness ⏥
b. Straightness —
c. Roundness ○
d. Cylindricity ⌭
e. Profile of a Line ⌒
f. Profile of a Surface ⌓
g. Parallelism ∥
h. Perpendicularity ⟂
i. Angularity ∠
J. Feature Control Frame | ⌖ | ⌀0.5Ⓜ | A | BⓂ | C |
k. Position ⌖
l. Runout (Circular) ↗
m. Concentricity ◎
n. Runout (Total) ⌰

15) Identify the following symbols:

a. Ⓜ = **maximum material condition**
b. ↧ = **depth**
c. ⌀ = **diameter**
d. Ⓟ = **projected tolerance zone**
e. ⟨ST⟩ = **statistical tolerance**
f. Ⓛ = **least material condition**
g. ⌴ = **counterbore**

5. Datums, Datum Features and Datum feature Simulators

1) What geometrically perfect elements are representative of datums? **points, axes and planes**

2) How are datum features simulated in manufacturing or inspection? **with datum feature simulators, such as surface plates, angle plates and chucks**

3) What criteria are used to choose a datum feature when tolerancing? **Functionality, features that are representative of seating, mating and alignment features, their accessibility and repeatability.**

4) What is a datum reference frame and what is it used for? **Three mutually perpendicular planes. A coordinate system is used to orient to and from which to measure angle and distance/location.**

5) What six degrees of spatial freedom does a part have before it is restrained? **A part must have three linear degrees of freedom (x, y and z) and three rotational degrees of freedom (u, v and w).**

6) **a minimum of three**; **a minimum of two**; **a minimum of one**

7) Are three datums always necessary for part orientation? If yes, why? If no, why not? **No. Some relationships require less that full part stabilization. Coaxiality and symmetry are two simple examples of this.**

8) Is a datum feature perfect? **No. A datum feature is real and therefore flawed.**

9) Is a datum feature ever given form control of its own? If not, why is it not necessary? If so, why it is necessary? **Yes, it is necessary to make it more repeatable and better shaped for part stabilization.**

10) How is the order of precedence of datum features shown in a feature control frame? **Read from left to right with the primary first, etc. It has nothing to do with the order of the letters as they appear in the alphabet. It has only to do with the order in which they are referenced.**

11) Why does the order of precedence of datum features make such a difference in feature orientation? **The primary dictates the angle of the secondary. The primary and secondary dictate the angle of the tertiary.**

12a) **with a minimum circumscribed cylinder (actual mating envelope)**
12b) **with a maximum inscribed cylinder (actual mating envelope)**

13) Of what use is an angular orientation datum in a datum reference frame when it is referenced after a cylindrical datum feature? **It orients the datums that have preceded it. It can be used to stop pattern rotation and keep the desired angular relationship between the pattern and the datums.**

14) How many datum planes are derived from a cylindrical datum feature? **two**

15) Why are datum targets necessary? **They are necessary to stabilize parts that are rough or warped or irregularly shaped or to duplicate partial areas of mating part contact.**

16) How are cylindrical or width datum features controlled Regardless of Material Boundary simulated? **by contacting the surface to establish an actual mating envelope axis or centerplane**

17) At what size or condition are secondary datum features of size that have their own tolerance of orientation or position simulated as datum features? **At their applicable virtual condition, if they are referenced at MMB: at their actual properly oriented mating size, if referenced at RFB.**

18) At what size is a primary datum feature of size referenced at Ⓜ represented for the features controlled to it? **At its MMC (which will be its applicable maximum material boundary)-if referenced at MMB (unless a straightness control generates a virtual condition boundary-MMC/MMB concept); at actual mating size if referenced RMB.**

19) What does a datum feature such as A(M)-B(M) mean? **It is when more than one feature such as a diameter, width or surface is used to establish a single axis or plane.**

20) When a pattern of holes is controlled to a datum feature hole, as the datum feature hole departs from MMB within its limits, what happens to the controlled hole pattern and how is it calculated? **The pattern of holes is allowed an apparent shift from the datum feature axis. Usually, this shift is one for one. Actually, the datum feature axis (actual feature axis) is allowed to shift away from the datum axis (imaginary datum axis). This appears as if the pattern of holes has shifted away from the datum feature axis.**

21) Do screw threads make good datum features? **Generally, they should be avoided because they generate a datum axis from the pitch diameter (also known as a pitch cylinder) that is hard to simulate in inspection. Sometimes however, when no other functional feature is appropriate, they must be used.**

22) If more than one datum reference frame is specified for a part, what effect will this have on manufacturing and inspection? **It implies a different set-up for each datum reference frame.**

23) Why would a designer choose to use only a portion of a surface as a datum feature instead of the entire surface? **Perhaps only a portion of the surface is functional or contacts the other parts in the assembly.**

24) In what forms are datum targets specified? **points, lines and areas**

25) If a designer wants to show a datum target as originating from the opposite side of a part than the view it has been shown in, how may it be done without another view? **with a dashed radial leader line that terminates at the datum target**

26) What kind of contact is required from the physical datum feature simulation equipment, in a datum target area, on a non-cylindrical surface? **Full flat contact with the entire area is desired. High point contact within the area is attempted.**

27) How is the shape and location of a datum target area defined? **either by basic or toleranced dimensions and sometimes through the use of chain lines and hatched lines**

28) What does the dimension origin symbol look like and of what use is it? **It is a circle used in place of an arrowhead on one end of a dimension line, and it designates that surface as generating an origin plane for measurement.**
Example:

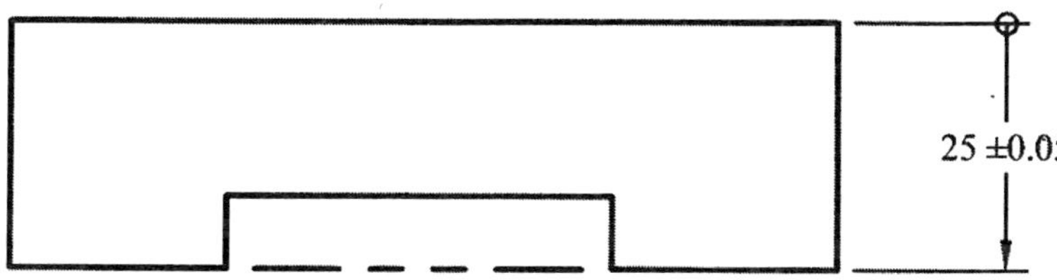

6. Virtual Condition and Resultant Condition Boundaries

1) When a feature has errors of orientation, location and certain tolerances of form, its effective mating boundary can be virtually altered at that orientation or at that location or by its form.
True

2) In virtual condition (MMC concept), the effective mating boundary of a hole is virtually increased and the shaft size virtually decreased, even though they still measure within their limits of size.
False. The hole is virtually decreased and the shaft is virtually increased.

3) A virtual condition boundary (MMC concept) of the feature can represent the worst condition of assembly of the mating components at a particular orientation or location.
True

4) If a feature is allowed a virtual condition, it may exceed its limits of size as an individual feature when measured.
False

5) How does calculating the virtual condition of an external feature of size differ from calculating the virtual condition (MMC concept) of an internal feature of size?
For holes: MMC minus the geometric tolerance applicable equals the virtual condition.
For shafts: MMC plus the geometric tolerance applicable equals the virtual condition.

6) Functional gages are made to the virtual condition (MMC concept) of features they inspect.
True

7) Functional gages are usually only made to check features controlled with MMC. Otherwise, for checking features controlled RFS, a functional gage would have to be made to take into account every size the feature could be produced at.
True

7. Form Controls

FLATNESS

1) Describe the configuration of a flatness tolerance zone? **two parallel planes separated by the geometric tolerance**

2) Is flatness considered individual or related to datums? **individual**

3) Is a datum feature reference allowable in the feature control frame? **No**

4) May a diameter sign be included in the feature control frame? **No**

5) Is flatness a surface or an axis control? **a surface control**

6) Is flatness a two-dimensional or a three-dimensional control? **three-dimensional**

7) To obtain a high level of certainty (for example: 90% certainty) of how flat a surface is, when measuring flatness will a few points from the surface suffice, or does this require hundreds of points per square inch? **Approximately 191 points per square inch (between 80 and 357 points) for a 90% certainty of flatness according to a study done by Electronic Measuring Devices, Inc. They also said that for a 75%**

certainty an average of 102 points (between 48 and 227 points) per square inch was required. For their study (done in 1988) they scanned in 2500 evenly spaced points on 6 one-inch blocks to represent 100% certainty of flatness. Each block had a different deformity (convex, concave, gouged, etc.)

8) If a surface is controlled with flatness, may the controlled surface in combination with its opposing side on the part violate an envelope of perfect form at MMC? **No**

9) Must the flatness tolerance refine the size tolerance or may it be larger? **It must refine (be smaller than) the size tolerance.** Why? **If the size tolerance was tighter than the flatness tolerance, then the flatness of the surface would be controlled better by the size tolerance (and the flatness tolerance would be meaningless).**

10) May a MMC symbol be included in the feature control frame if flatness is applied to a single surface? **No**

11) How is the flatness control directed to the controlled surface? **by a leader line to the surface or an extension line off of the surface**

12) May flatness be applied on a unit or a rate basis? **Yes**

13) May flatness be applied to control a derived median plane per ASME Y14.5-2009? **Yes**

STRAIGHTNESS

General:

1) Are datum feature references allowable/required in the feature control frame? **No, they are not allowable.**

Surface Straightness (question 2 through 8):

2) What view should the straightness of a surface tolerance be applied in? **For a surface control, it is shown in a view that the tolerance zones run parallel to. Straightness of a surface applies only in the plane of the view in which it is shown.**

3) In straightness of a surface, what is the shape of the tolerance zone, and what must lie within it if the feature is to be accepted? **Two parallel lines that are separated by the geometric tolerance. A separate tolerance zone applies to each line element. Each tolerance zone may be on a different height or on a different angle to the other tolerance zones. Each point on the line under test must reside within its tolerance zone.**

4) What material condition symbols may be included in the feature control frame? **none for surface straightness**

5) May a diameter sign appear in the feature control frame? **not for surface straightness**

6) Without a straightness feature control frame, what controls the straightness of a part surface? **The limits of size. The part must have perfect form if produced at MMC (if the feature is rigid and no other notes or controls have eliminated the requirement that size tolerance controls form).**

7) Is perfect form of the controlled surface and its opposing surface required at MMC? **Yes**

8) Is bonus tolerance ever possible? **not for surface straightness** If not, why not? **No surface control is allowed to use a maximum material condition symbol after the geometric tolerance.**

Straightness of the Derived Median Line (questions 9 through 15) :

9) In straightness of a derived median line, what is the tolerance zone shape and what must be contained within it for the feature to be acceptable? **The tolerance zone is cylindrical in shape. The derived median line must reside within this tolerance zone. The derived median line is an imperfect line (abstract) that passes through the center points of all cross sections of the feature. These cross sections are normal to the axis of the actual mating envelope.**

10) Is perfect form of the controlled feature required at MMC? **No. Perfect form at MMC is no longer required when this control is used.**

11) Is this control considered individual or related to datums? **individual**

12) How might straightness of a derived median line that is controlled with a MMC symbol be gaged? **with a functional gage made at the virtual condition of the feature under test**

13) How might straightness of a derived median line that is controlled with the RFS concept be measured? **with differential measurements to determine the derived median line as it is described in Question 9**

14) Is a virtual condition allowed for both controls described in Questions 12 and 13? **No. A virtual condition is defined as a constant boundary. A constant boundary is created that can be gaged only for features controlled at MMC, not at RFS.**

15) Must the straightness tolerance be smaller than the feature's size tolerance? **No. For straightness of the derived median line, the geometric tolerance may be smaller or larger than the size tolerance.**

CIRCULARITY

1) Is circularity a two- or a three-dimensional control? **two-dimensional**

2) What is the tolerance zone shape? **The tolerance zone shape is two concentric circles that are the geometric tolerance apart radially. Each circular cross-section has its own tolerance zone. Each tolerance zone may have a different optimal center offset from the others. Although the zones are the same distance apart (the size of the circularity tolerance), the diameters of the zones are determined optimally by the circles at each cross-section.**

3) What material condition symbols are allowed in the feature control frame? **none**

4) Are datum feature references allowable? **No**

5) Is a feature controlled with circularity allowed to exceed its size limits? **No**

6) Is any extra tolerance derived from the controlled feature's size limits as the feature departs from MMC? **No**

7) How is circularity inspected if done with a dial indicator? **The full indicator movement in a 360° revolution for each circular cross-section tested may not exceed the circularity tolerance in the feature control frame. The inspector is responsible for every circular cross-section of the surface being controlled.**

8) Does the tolerance in the feature control frame have to be smaller than the feature's size limits? **Yes**

9) Is circularity considered individual or related to datums? **individual**

10) Can circularity controls be assigned to cones? **Yes**

11) May a diamctcr sign be included in the circularity feature control frame? **No**

12) How is the feature control frame connected to the controlled surface? **with a leader line to the surface**

13) Circularity controls the surface **b. at every cross section.**

14) Without a circularity feature control frame, what would normally control a feature's roundness? **the size tolerance**

15) A circularity control is assumed to apply **b. RFS.**

CYLINDRICITY

1) Name at least two things cylindricity controls that circularity does not. **Cylindricity controls straightness and taper. Circularity does not.**

2) List three different ways cylindricity can be checked. Order them with the best way first and the least desirable method last.
   1. **Mounted in a roundness checking machine while points are collected and graphed while the part rotates and the electronic probe moves across (or up, depending on the orientation of the machine) the surface.**
   2. **Mounted between centers and a full indicator movement taken while the part rotates quickly and the dial indicator probe moves slowly down the surface**
   3. **With a coordinate measuring machine with a constant contact or non-contact scanning probe.**

3) What is the tolerance zone configuration a cylindricity control generates? **two concentric cylinders separated radially by the amount specified in the feature control frame**

4) Is cylindricity a surface or an axis control? **a three-dimensional surface control**

5) May a least material condition symbol be used in the feature control frame? **No**

6) Can cylindricity and circularity be applied to the same surface? If so, which tolerance must be smaller and why? **Yes they can, if the circularity tolerance is smaller because cylindricity controls circularity, straightness and taper.**

7) Does a cylindricity tolerance ever reference datum features inside the feature control frame? **No**

8) Must the controlled feature, if produced at MMC, have perfect form? **Yes**

9) Is the cylindricity tolerance additive to the controlled feature's MMC size limit? **No**

10) Is bonus tolerance ever allowed? **No**

11) A cylindricity tolerance is considered **a. form.**

12) A produced feature's cylindricity is compared to a perfect geometric counterpart of **c. itself.**

13) List all items allowed in the cylindricity feature control frame. **a geometric characteristic symbol (cylindricity) and a geometric tolerance**

14) A cylindricity tolerance controls roundness and
c. straightness and taper.

15) The tolerance applies simultaneously to circular and ____ elements of the controlled surface.
a. longitudinal

8. Orientation Controls

1) What is the configuration of the tolerance zone defined by a parallelism of a planar surface control? **The configuration is two parallel planes that are parallel to the datum. The two parallel planes are separated by the geometric (parallelism) tolerance.**

2) What does that tolerance zone take its orientation from? **the datum**

3) If we wanted an orientation tolerance zone | ∠ | 0.5 | for a planar surface to be exactly 35° from datum plane R as well as perfectly perpendicular to datum plane S, what could we do to the given feature control frame per ASME Y14.5-2009 to reflect that? | ∠ | 0.5 | R | S | **and make the 35° a basic angle.**

4) If a part has a maximum height of 1.000. Is a .002 parallelism control additive to that height? **No**

5) Is at least one datum feature reference always required in the feature control frame for angularity? **Yes**

6) How does the tolerance configuration for the angularity control in Question 3 differ from the tolerance zone that would be generated if the 35° was ± 2° ? **The zone for Question 3 consists of two parallel planes that are 0.5 apart and oriented 35° from datum R and 90° from datum S. This tolerance zone of ± 2° consists of a wedge-shaped zone originating at a vertex allowing the surface to vary a total of 4°.**

7) What material condition symbols would be appropriate in feature control frames for orientation controls if they were used on a single planar surface? **None would be appropriate after the orientation geometric tolerance, but either Ⓜ or Ⓛ would be after any datum features of size that are referenced.**

8) What is the usual configuration and orientation of a controlled feature's perpendicularity tolerance zone if it is used on a planar surface? **Two parallel planes separated by the perpendicularity tolerance. Both planes are 90° to the datum or datums referenced.**

9) What does the use of a datum feature reference add to the orientation geometric control? **It shows to what the desired relationship is to be held.**

10) ∅.250±.005 Hole
| ∠ | ∅.005Ⓜ | A |

If the controlled hole was produced at a diameter of .250, what angularity tolerance would apply? **∅.010**

11) Does the controlled hole in Question 10 have a virtual condition? If so, what is it? **The controlled hole in Question 10 has a virtual condition of ∅.240.**

12) Would it be appropriate to include a maximum material boundary symbol in the feature control frame for datums that are planar surfaces? Please explain. **No. Planar surfaces don't have a maximum material condition. Only features of size have a maximum material condition.**

13. How many contact points are required to establish datum A from a planar surface if A is primary? **3 high points of contact are required.**

14) Is an orientation geometric control related to datums or individual? **always related to datums**

15) For parallelism used on a planar surface to a datum plane generated by an opposing planar surface, must the tolerance in the feature control frame be smaller than the size tolerance of the part's opposing sides that are included in the control? **Yes. If the size tolerance was smaller, then it would control parallelism better than the parallelism control would and the parallelism control would be meaningless.**

16) ∅25±0.5 Hole

⊥	∅0.5Ⓜ	AⓂ	B

If the controlled hole departs from MMC to LMC, how much **extra** perpendicularity tolerance can be drawn from the size? **1 millimeter**

17) List the MMC and LMC of the datum feature in Question 16. **MMC=24.5 and LMC=25.5**

18) If the datum feature A referenced in question 16 is a hole and has a size tolerance of ∅50±0.1, as it departs from MMB to LMC, how much additional shift of the controlled hole's tolerance zone is gained? **∅0.2**

19) If both the controlled hole and datum feature A were made at LMC, how much could their axes be out of perpendicularity to each other? **1 + 0.5 + 0.2 = ∅1.7**

20) What is the shape of the tolerance zone generated by this geometric control? **cylindrical tolerance zone that is oriented to datum axis A and datum B that must contain the axis of the hole**

9. Profile

1) In a profile of a curved surface control, the profile must be dimensioned with **d. basic dimensions.**

2) What is the difference between profile of a line and profile of a surface? **Profile of a line is a two-dimensional line element control. Profile of a surface is a three-dimensional surface control.**

3) For profile controls, datum feature references are **a. allowable but not required.**

4) If a datum feature reference was used for a profile control, what purpose would be served? **to orient or locate (or both locate and orient) the tolerance zone**

5) Are material condition symbols allowed for the feature being controlled? Why or why not? **No. Profile is not allowed to use MMC or LMC symbols after the geometric tolerances. However, MMC or LMC symbols may be used after any datum features of size that are referenced in the control.**

6) If both profile of a surface and profile of a line are used on the same feature, which must have the smaller profile tolerance? **Profile of a line must be smaller.**

7) What is the tolerance zone shape of profile of a line? **two uniform lines separated by the geometric tolerance**

8) What is the tolerance zone shape of profile of a surface? **two uniform boundaries separated by the geometric tolerance**

9) How is profile tolerance designated as unequal-bilateral or unilateral? **If phantom lines are shown, they designate a direction of tolerance. The symbol for unilateral or unequal is preferred per Y14.5-2009.**

10) How is the extent of where the profile control applies on a surface designated? **with notes, between symbols, all-around symbols or the ALL OVER symbol or note**

11) Suppose the entire outer periphery of the part profile must be controlled. How may it be done? **It may be done wither with an all-around symbol or with an "ALL OVER" note per Y14.5-1994 or an "ALL OVER" symbol per Y14.5-2009 . All-around means that the control is view dependant. ALL OVER means that the control applies in all views.**

10. Runout

1) a
2) b
3) d
4) a
5) For cylindrical surfaces: roundness, straightness, taper, coaxiality, angularity and surface profile (or cylindricity and concentricity).
   For flat surfaces: flatness and perpendicularity to the datum axis.
6) Roundness/circularity and coaxiality for circular features.
7) When balance is important but 3-D form is not as important.
8) d
9) Related
10) Required
11) No
12) No
13) Yes
14) No
15) Circular runout must be smaller

11. Location

1) c
2) a
3) No
4) Yes..
5) A diameter. Generally, it is a cylindrical tolerance zone that must contain the median points.
6) Related
7) c
8) a
9) c. When balance for a rotating feature is important, surface form is not important.
10) No
11) No
12) b
13) a
14) No
15) No
16) Two parallel planes
17) Yes
18) e
19) b
20) Yes
21) a. basic dimensions, datum features and a feature control frame.
22) Three mutually perpendicular planes
23) Datum features
24) The form of a datum feature being irregular
25) Because basic dimensions are used, there is no tolerance accumulation.
26) The full datum reference frame serves as the origin of measurement for location and orientation and stabilizes the part's six spatial degrees of freedom.
27) A position symbol, perhaps a diameter sign or a spherical diameter sign, a geometric tolerance, specified or implied material condition symbols, perhaps a projected tolerance zone symbol and a height for the projected tolerance zone, perhaps datum feature references and perhaps material condition boundary symbols.
28) Size tolerance
29) A floating fastener assembly consists of two parts with clearance holes in each. A fastener floats through holes in both parts, and a nut binds the parts together.
30)

MMC Hole	Geo. Tol. at MMC	MMC Hole
-MMC Fastener	+MMC Fastener	-Geo. Tol. at MMC
Geo. Tol. for All Holes on Both Parts	MMC Hole	MMC Fastener

31) a) .265 - .250 = .015
    b) .520 - .500 = .020
32) a) .775 - .025 = .750
    b) 1.030 - .030 = 1.000

33) a) .3125 + .010 = .3225
b) .625 + .020 = .645

34. The maximum height of the screw body or stud that protrudes from the threaded hole. This sometimes equates to the maximum thickness of the mating part.

35) Because of variations in orientation of the threaded holes pitch diameter and how this is magnified over the height of the screw when the screw is threaded in

36) Threaded holes and tight fitting holes, such as dowel holes, which dictate the angle of the screw or dowel

37)

MMC Hole	Virtual Cond. Hole	MMC Hole
- MMC Screw	- MMC of Shaft or Screw	- Virt. Cond. Shaft
Geo. tol. divided bet. parts	Geo. Tol. Shaft	Geo. Tol. Hole

12. General Rules, Formulas and Measurement Questions

1) Unless otherwise specified, limits of size are measured in the following manner:
a) LMC at cross sections and MMC for violations of a full form envelope.

2) When basic dimensions are given to locate a hole,
c) the location tolerance is found in a position feature control frame.

3) The correct formula for 'bonus tolerance' for a hole controlled with the MMC concept is
d) actual size minus MMC.

4) Total runout can be inspected the best with
c) a dial indicator.

5) Position deviation from 'true position' can be calculated using the following formula:
a) two times the square root of the sum of the squares of the x and y deviations.

6) A coordinate measurement machine can most accurately simulate a primary datum by
c) probing the surface plate on which the highest points of the datum feature rest.

7) The datum axis generated by an external cylindrical datum feature is the axis of the
a) minimum circumscribed cylinder.

8) An optical comparator is most often associated with the inspection of
b) profile.

9) Rocking datum features may
d) be adjusted to an optimum position.

10) A cylindricity geometric characteristic symbol with a geometric tolerance of .003 allows
d) size variations for the cylinder's diameter as large as .006.

11) Which of the following would be inappropriate to include in a feature control frame for a straightness control?
c) a datum reference

12) Runout is what type of control?
a) surface to datum axis

13) Circular runout of a diameter does not control which of the following characteristics?
c) straightness of the surface

14) Position is a geometric control that is not capable of controlling
d) circularity

15) Of the following geometric characteristic symbols, which is the most restrictive in its control of surface characteristics?
c) total runout

16) Of the following geometric characteristic symbols, which is considered a 2-D line element control?
a) circularity

17) Of the following geometric characteristic symbols, which is least capable of controlling flatness?
d) straightness

18) Limits of size on a rectangular part are not a control of which of the following?
b) perpendicularity

19) Virtual condition (MMC concept) is calculated for a hole by the following formula:
a) MMC of the hole minus the geometric tolerance applicable at MMC.

20) Positional tolerance for all holes in a floating fastener assembly can be calculated in the following manner:
b) MMC of the hole minus the MMC of the mating fastener.

21) What is the virtual condition of the following shaft?
∅8.3 – 8.5

⌖	∅0.3Ⓜ	A	B	C

c) ∅8.8

22) When a pattern of holes is positioned to a datum feature of size such as a hole and this datum feature is referenced at MMB, as the datum feature (hole) grows,
b) the pattern of holes as a group may shift.

23) In the concept of composite positional tolerancing,
a) the features within a pattern hold a tighter tolerance in relation to one another than to any datums that are listed only in the upper level control.

24) When a pattern of holes is used to establish a datum reference frame from which to locate other part features,
d) an infinite number of holes may be referenced to form the datum reference frame.

25) When multiple patterns of features are located from the same datum reference frame, they are considered to be
a) a simultaneous requirement.

26) If a pattern of holes is positioned and no datum features are used in their positional feature control frame,
b) the holes are considered positioned to one another.

27) Two hole patterns are positioned to the same primary planar datum feature, and this datum feature, due to imperfections of flatness, tends to rock. If the patterns of holes are considered as separate requirements,
a) the part may be rocked one way to accept one pattern of holes and another way to accept the other pattern of holes during inspection.

28) A four-hole pattern has size limits of ∅10.7 – 11.0 and a position tolerance to a datum reference frame of ∅0.3 at MMC. Its mating pin pattern has size limits of ∅9.7 – 10.0 and is positioned to a representative set of datums. How much position tolerance is available for the pins within the pin pattern at MMC?
d) ∅0.4

29) For the circumstances given in Question 28, the position tolerance that would be allowed for a hole produced at ∅11 is
c) ∅0.6

30) For the circumstances given in Question 28, what is the virtual condition (MMC concept) for the pins within the pin pattern?
d. ∅10.4

Holes:

1.	∅.625 ±.010	Ⓜ	.615
		Ⓛ	.635
2.	∅ .620 / .590	Ⓜ	.590
		Ⓛ	.620
3.	∅.835 ±.005	Ⓜ	.830
		Ⓛ	.840
4.	∅ .705 / .695	Ⓜ	.695
		Ⓛ	.705
5.	∅ 1.265 / 1.245	Ⓜ	1.245
		Ⓛ	1.265

Shafts:

6.	∅ .810 / .750	Ⓜ	.810
		Ⓛ	.750
7.	∅.710 ±.020	Ⓜ	.730
		Ⓛ	.690
8.	∅ 1.265 / 1.245	Ⓜ	1.265
		Ⓛ	1.245
9.	∅ 22.875 / 21.885	Ⓜ	22.875
		Ⓛ	21.885
10.	∅ .090 / .085	Ⓜ	.090
		Ⓛ	.085

Answer to page 539

Answer to page 540

1b. Calculate the MMC and LMC for the following holes and shafts.

Holes:

1. **MMC= 62.4 LMC= 62.6**

2. **MMC= 59 LMC= 62**

3. **MMC= 82.45 LMC= 83.55**

4. **MMC= 69.5 LMC= 70.5**

5. **MMC= 1.245 LMC= 1.265**

Shafts:

6. **MMC= 8.1 LMC= 7.5**

7. **MMC= 71.2 LMC= 70.8**

8. **MMC= 12.65 LMC= 12.45**

9. **MMC= 22.875 LMC= 21.885**

10. **MMC= 0.90 LMC= 0.85**

ANSWER to page 541

Exercise 2 Section 2

	Feature / Feature Control Frame	Outer Boundary	Inner Boundary
1)	Ø20±2 Hole [⌖][Ø 2 Ⓜ][A][B][C]	22+6=28	18−2=16
2)	Ø20±2 Hole [⌖][Ø 2 Ⓛ][A][B][C]	22+2=24	18−6=12
3)	Ø20±2 Hole [⌖][Ø 2][A][B][C]	22+2=24	18−2=16
4)	Ø20 $^{+3}_{-2}$ Shaft [⌖][Ø 0.5 Ⓜ][A][B][C]	23.0+0.5=23.5	18.0−5.5=12.5
5)	Ø15−16 Shaft [⌖][Ø 0.4 Ⓛ][A][B][C]	16.0+1.4=17.4	15.0−0.4=14.6
6)	Ø30 $^{+0.4}_{-0.2}$ Shaft [⌖][Ø 0.3][A][B][C]	30.4+0.3=30.7	29.8−0.3=29.5
7)	28.5 $^{+0.2}_{-0.1}$ Slot [⌖][0.8 Ⓜ][X][Y][Z]	28.7+1.1=29.8	28.4−0.8=27.6
8)	23.08−24.04 Tab [⌖][0.5 Ⓜ][T][E][C]	24.04+0.50=24.54	23.08−1.46=21.62
9)	75±2 Slot [⌖][1 Ⓛ][C][A][B]	77+1=78	73−5=68
10)	70±0.5 Tab [⌖][0.5 Ⓛ][D][E][F]	70.5+1.5=72	69.5−0.5=69
11)	12−13 Slot [⌖][0.3][G][A][F]	13+0.3=13.3	12−0.3=11.7
12)	9.5−10.6 Tab [⌖][1.2][H][J][K]	10.6+1.2=11.8	9.5−1.2=8.3

①

$$\begin{array}{r} 28 = \text{O.B.} \\ +\ 16 = \text{I.B.} \\ \hline 44 = \text{Sum} \end{array} \qquad \begin{array}{r} 28 = \text{O.B.} \\ -\ 16 = \text{I.B.} \\ \hline 12 = \text{Diff.} \end{array}$$

$$\frac{44}{2} = 22 = \text{Mean} \quad \& \quad \frac{12}{2} = 6 = \pm\text{Tol.}$$

So; 22±6

②

$$\begin{array}{r} 24 = \text{O.B.} \\ +\ 12 = \text{I.B.} \\ \hline 36 = \text{Sum} \end{array} \qquad \begin{array}{r} 24 = \text{O.B.} \\ -\ 12 = \text{I.B.} \\ \hline 12 = \text{Diff.} \end{array}$$

$$\frac{36}{2} = 18 = \text{Mean} \quad \& \quad \frac{12}{2} = 6 = \pm\text{Tol.}$$

So; 18±6

③

$$\begin{array}{r} 24 = \text{O.B.} \\ +\ 16 = \text{I.B.} \\ \hline 40 = \text{Sum} \end{array} \qquad \begin{array}{r} 24 = \text{O.B.} \\ -\ 16 = \text{I.B.} \\ \hline 8 = \text{Diff.} \end{array}$$

$$\frac{40}{2} = 20 = \text{Mean} \quad \& \quad \frac{8}{2} = 4 = \pm\text{Tol.}$$

So; 20±4

④

$$\begin{array}{r} 23.5 = \text{O.B.} \\ +\ 12.5 = \text{I.B.} \\ \hline 36\ \ = \text{Sum} \end{array} \qquad \begin{array}{r} 23.5 = \text{O.B.} \\ -\ 12.5 = \text{I.B.} \\ \hline 11\ \ = \text{Diff.} \end{array}$$

$$\frac{36}{2} = 18 = \text{Mean} \quad \& \quad \frac{11}{2} = 5.5 = \pm\text{Tol.}$$

So; 18±5.5

⑤

$$\begin{array}{r} 17.4 = \text{O.B.} \\ +\ 14.6 = \text{I.B.} \\ \hline 32\ \ = \text{Sum} \end{array} \qquad \begin{array}{r} 17.4 = \text{O.B.} \\ -\ 14.6 = \text{I.B.} \\ \hline 2.8 = \text{Diff.} \end{array}$$

$$\frac{32}{2} = 16 = \text{Mean} \quad \& \quad \frac{2.8}{2} = 1.4 = \pm\text{Tol.}$$

So; 16±1.4

⑥

$$\begin{array}{r} 30.7 = \text{O.B.} \\ +\ 29.5 = \text{I.B.} \\ \hline 60.2 = \text{Sum} \end{array} \qquad \begin{array}{r} 30.7 = \text{O.B.} \\ -\ 29.5 = \text{I.B.} \\ \hline 1.2 = \text{Diff.} \end{array}$$

$$\frac{60.2}{2} = 30.1 = \text{Mean} \quad \& \quad \frac{1.2}{2} = 0.6 = \pm\text{Tol.}$$

So; 30.1±0.6

⑦

$$\begin{array}{r} 29.8 = \text{O.B.} \\ +\ 27.6 = \text{I.B.} \\ \hline 57.4 = \text{Sum} \end{array} \qquad \begin{array}{r} 29.8 = \text{O.B.} \\ -\ 27.6 = \text{I.B.} \\ \hline 2.2 = \text{Diff.} \end{array}$$

$$\frac{57.4}{2} = 28.7 = \text{Mean} \quad \& \quad \frac{2.2}{2} = 1.1 = \pm\text{Tol.}$$

So; 28.7 ± 1.1

⑧

$$\begin{array}{r} 24.54 = \text{O.B.} \\ +\ 21.62 = \text{I.B.} \\ \hline 46.16 = \text{Sum} \end{array} \qquad \begin{array}{r} 24.54 = \text{O.B.} \\ -\ 21.62 = \text{I.B.} \\ \hline 2.92 = \text{Diff.} \end{array}$$

$$\frac{46.16}{2} = 23.08 = \text{Mean} \quad \& \quad \frac{2.92}{2} = 1.46 = \pm\text{Tol.}$$

So; 23.08 ± 1.46

⑨

$$\begin{array}{r} 78 = \text{O.B.} \\ +\ 68 = \text{I.B.} \\ \hline 146 = \text{Sum} \end{array} \qquad \begin{array}{r} 78 = \text{O.B.} \\ -\ 68 = \text{I.B.} \\ \hline 10 = \text{Diff.} \end{array}$$

$$\frac{146}{2} = 73 = \text{Mean} \quad \& \quad \frac{10}{2} = 5 = \pm\text{Tol.}$$

So; 73 ± 5

⑩

$$\begin{array}{r} 72 = \text{O.B.} \\ +\ 69 = \text{I.B.} \\ \hline 141 = \text{Sum} \end{array} \qquad \begin{array}{r} 72 = \text{O.B.} \\ -\ 69 = \text{I.B.} \\ \hline 3 = \text{Diff.} \end{array}$$

$$\frac{141}{2} = 70.5 = \text{Mean} \quad \& \quad \frac{3}{2} = 1.5 = \pm\text{Tol.}$$

So; 70.5 ± 1.5

⑪

$$\begin{array}{r} 13.3 = \text{O.B.} \\ +\ 11.7 = \text{I.B.} \\ \hline 25 = \text{Sum} \end{array} \qquad \begin{array}{r} 13.3 = \text{O.B.} \\ -\ 11.7 = \text{I.B.} \\ \hline 1.6 = \text{Diff.} \end{array}$$

$$\frac{25}{2} = 12.5 = \text{Mean} \quad \& \quad \frac{1.6}{2} = 0.8 = \pm\text{Tol.}$$

So; 12.5 ± 0.8

⑫

$$\begin{array}{r} 11.8 = \text{O.B.} \\ +\ 8.3 = \text{I.B.} \\ \hline 20.1 = \text{Sum} \end{array} \qquad \begin{array}{r} 11.8 = \text{O.B.} \\ -\ 8.3 = \text{I.B.} \\ \hline 3.5 = \text{Diff.} \end{array}$$

$$\frac{20.1}{2} = 10.05 = \text{Mean} \quad \& \quad \frac{3.5}{2} = 1.75 = \pm\text{Tol.}$$

So; 10.05 ± 1.75

Answer to page 542

Section 2-Basic Applications

3. Please convert the following mating parts from plus and minus location tolerancing to position tolerancing. For Parts #1 and #2:

Part #1

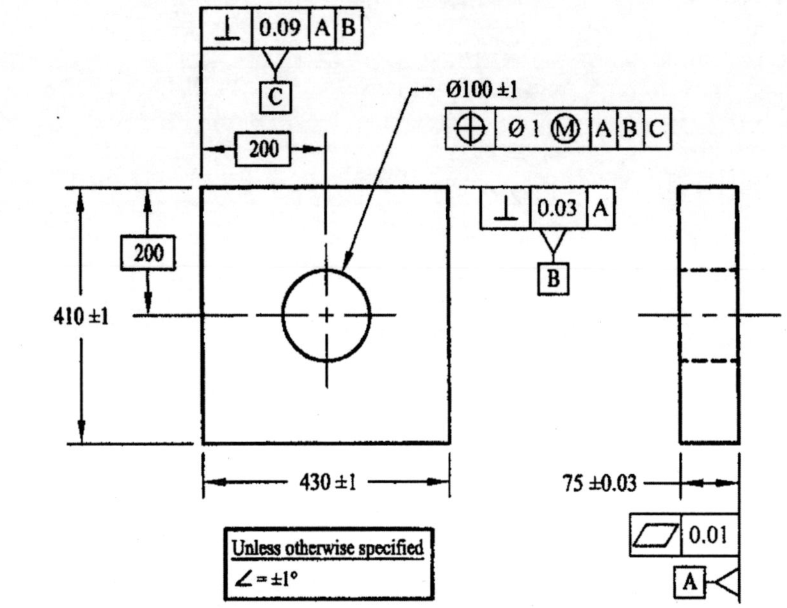

Part #2

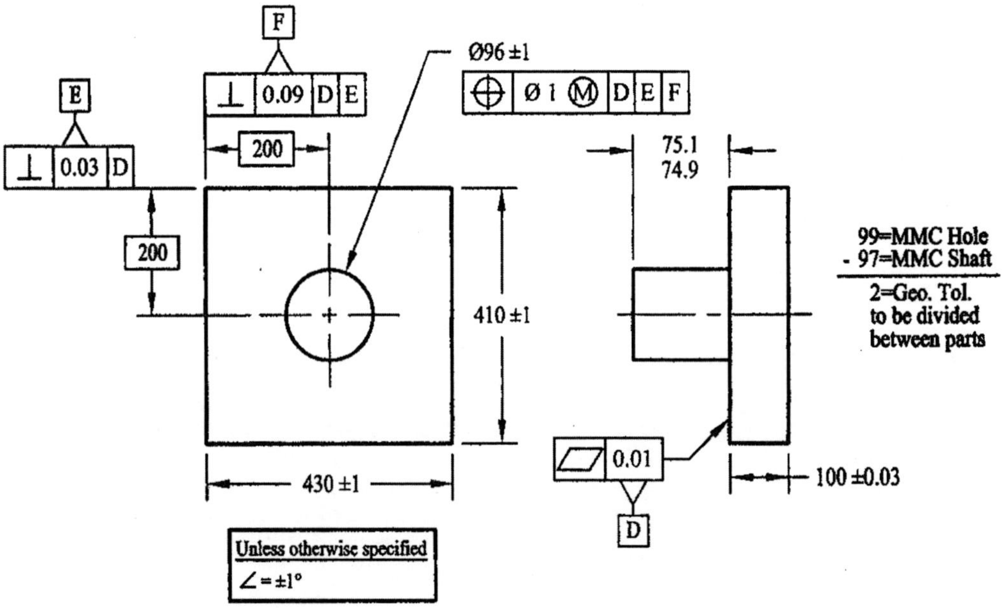

4.

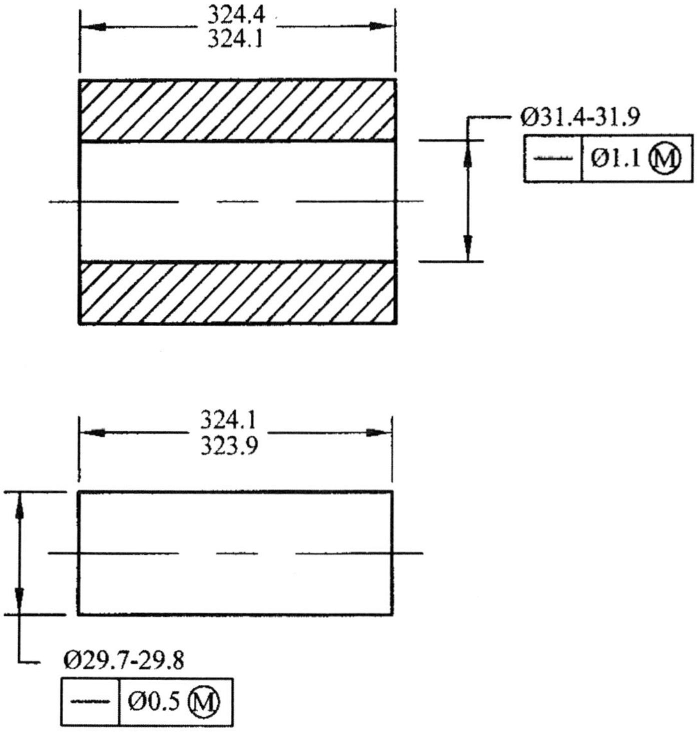

5.

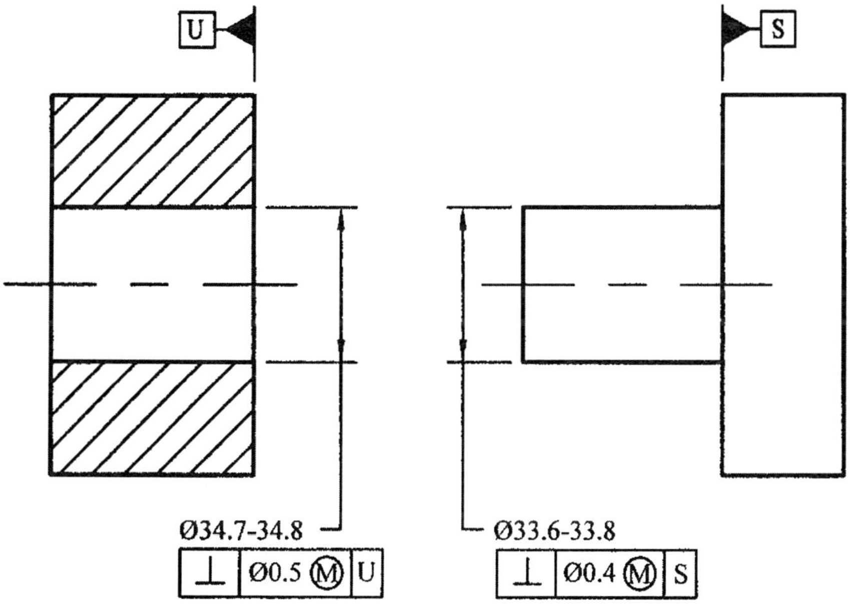

Answers to pages 543 and 544

6.

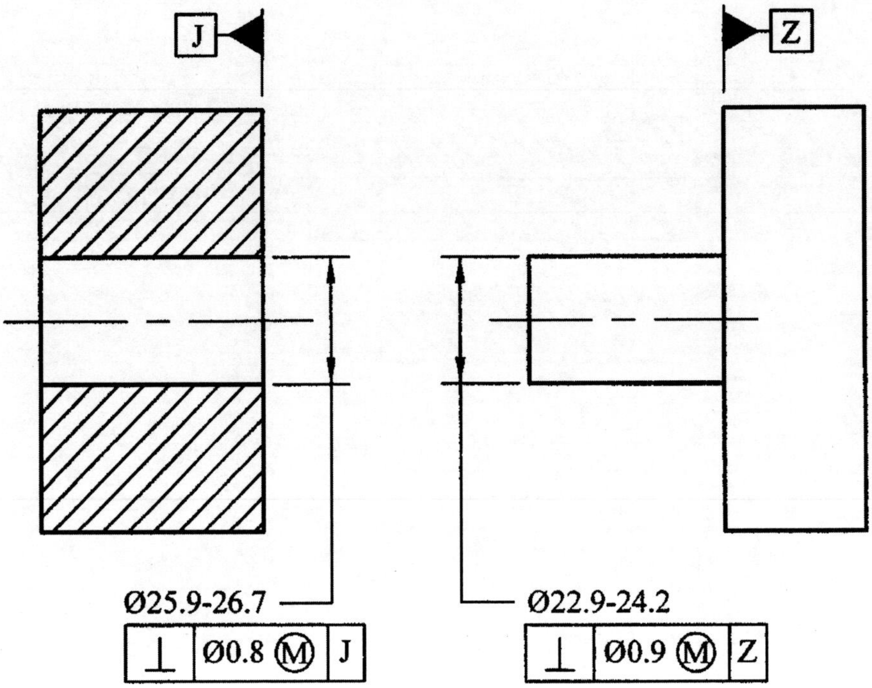

Answer to page 545

7.

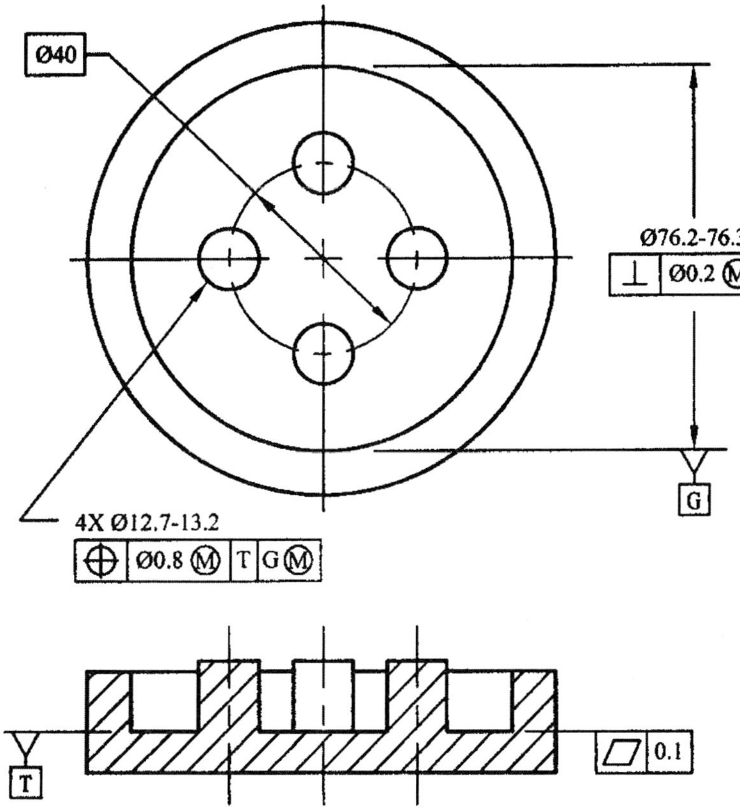

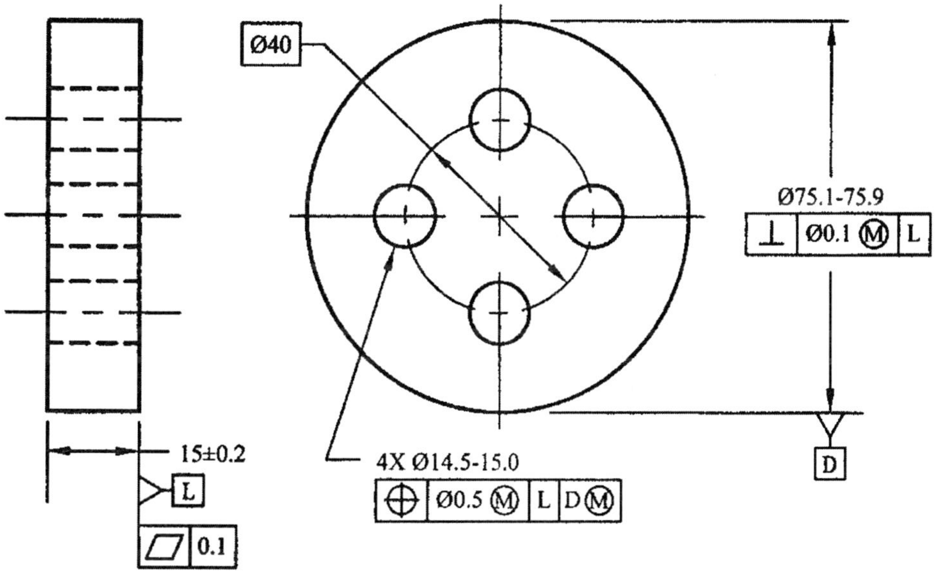

Answer to page 546

8.

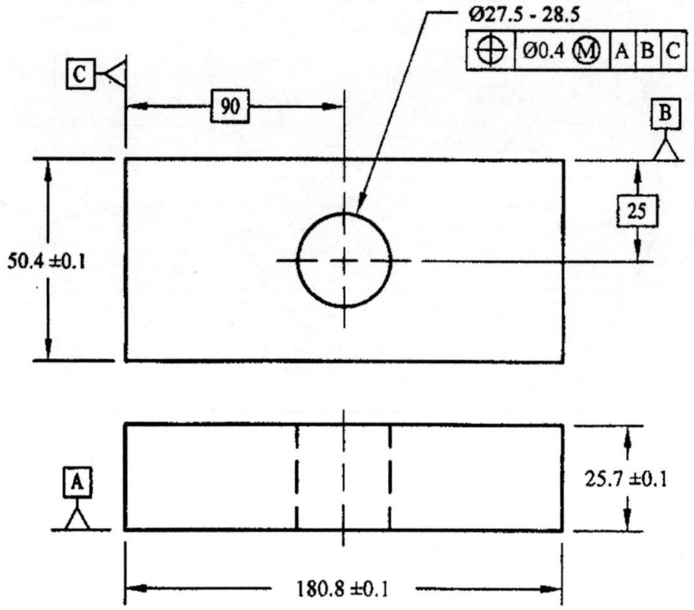

27.5 MMC
\- 0.4
① 27.1 Virt. Cond. to protect
② 27.75 protected
③ Good Part

AND

① 28.25 MIC
-27.50 MMC
0.75 Bonus Geo. Tol.
\+ 0.40 Orig. Geo. Tol.
Ø 1.15 Total Geo. Tol.

② 28.25 MIC
-27.75 MIC at ⊕
Ø 0.50 out of ⊕

③ Good Part

Part 1 as produced being inspected:

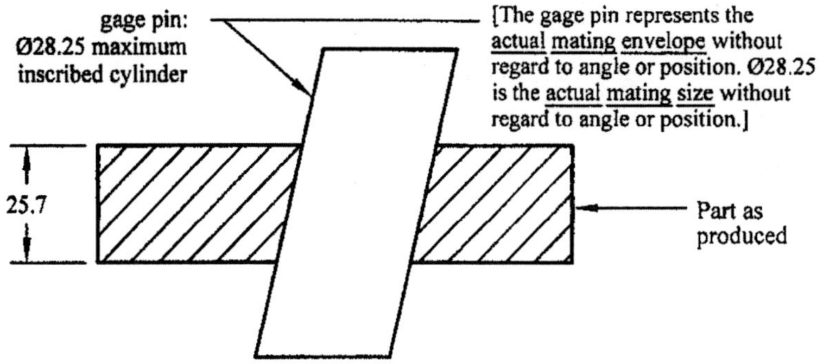

Part 1 as produced being inspected:

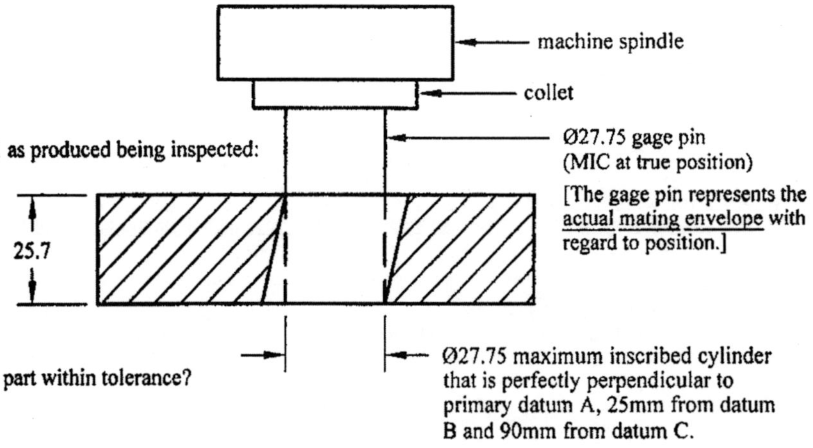

1. Is this part within tolerance?

Answer to page 547

9. **Answer to page 548**

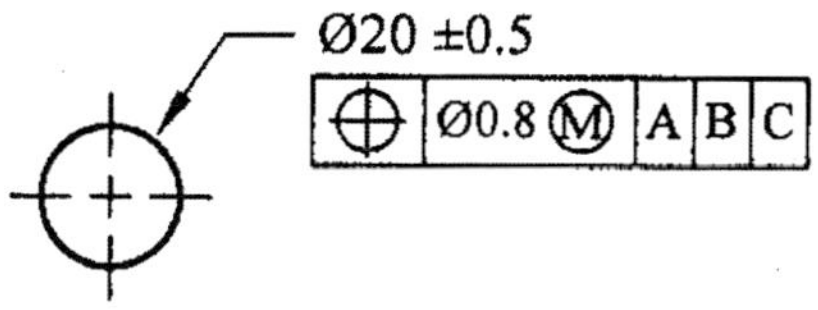

IF A HOLE

	Actual	Less MMC	Bonus Equals	Original Pos. Tol.	Total
A.	20.5	19.5	1	0.8	1.8
B.	20.3	19.5	0.8	0.8	1.6
C.	20.1	19.5	0.6	0.8	1.4
D.	20	19.5	0.5	0.8	1.3
E	19.8	19.5	0.3	0.8	1.1
F.	19.6	19.5	0.1	0.8	0.9
G.	19.5	19.5	0	0.8	0.8

IF A SHAFT

	MMC	Less Actual	Bonus Equals	Original Pos. Tol.	Total
H.	20.5	19.5	1	0.8	1.8
I.	20.5	19.6	0.9	0.8	1.7
J.	20.5	19.8	0.7	0.8	1.5
K.	20.5	20	0.5	0.8	1.3
L.	20.5	20.1	0.4	0.8	1.2
M.	20.5	20.3	0.2	0.8	1
N.	20.5	20.5	0	0.8	0.8

10. **Answers page 549**

Dimension from Datum C	Dimension from Datum B	Actual Hole Size	Answers		
			A	B	C
(1) 1.901	.924	.500	.015	.0028	YES
(2) 1.905	.929	.505	.020	.0128	YES
(3) 1.902	.926	.510	.025	.0045	YES
(4) 1.925	.932	.502	.017	.0519	NO
(5) 1.895	.922	.500	.015	.0117	YES
(6) 1.890	.925	.507	.022	.020	YES
(7) 1.900	.910	.509	.024	.030	NO
(8) 1.901	.926	.511	violates LMC	.0028	NO
(9) 1.870	.909	.509	.024	.068	NO

11. **Answers page 550**

as produced:

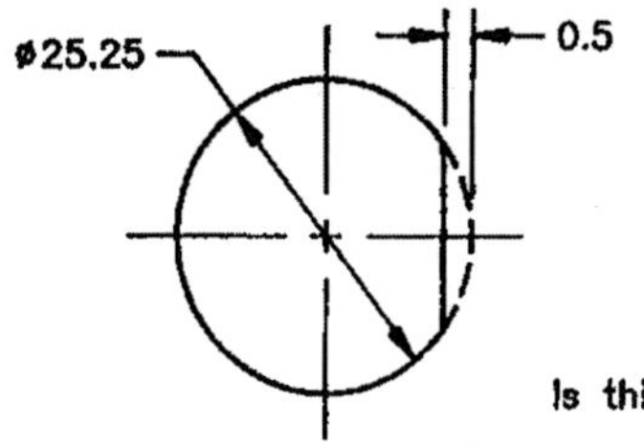

Is this part within tolerance ? **Yes**

as produced:

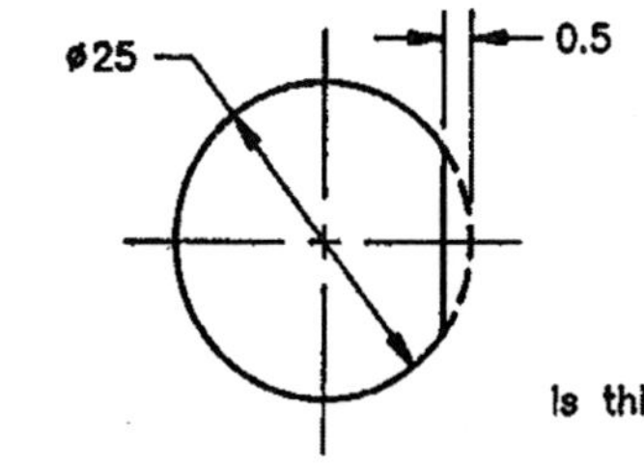

Is this part within tolerance ?

No – violates LMC

as produced:

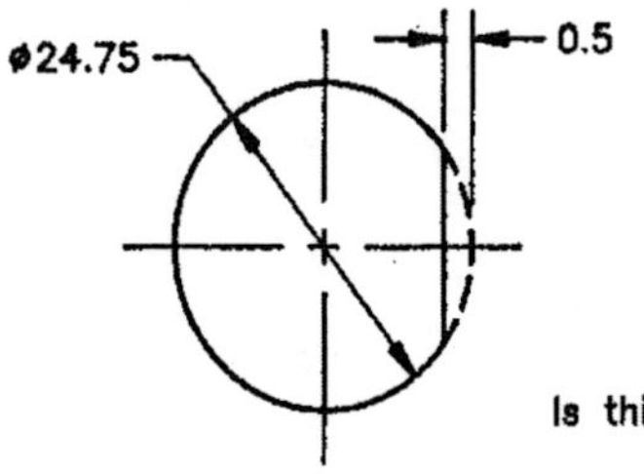

Is this part within tolerance ?

No – violates LMC

12. Answer to page 551 13. Answers page 552 14. Answers page 553

as produced:

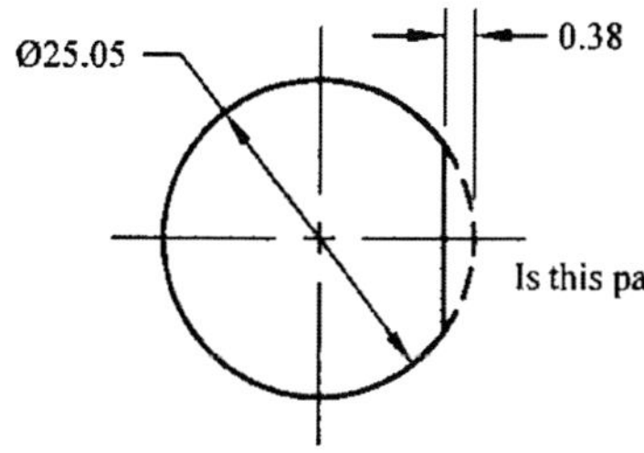

Is this part within tolerance ? No

Flat is deeper than
0.35 & violates LMC

as produced:

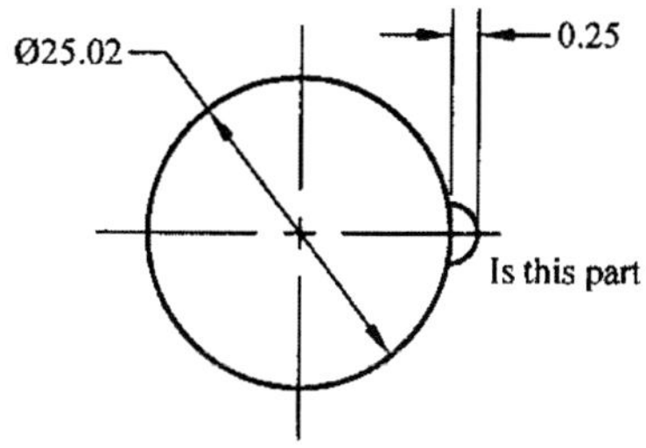

Is this part within tolerance ? No

Part is round enough,
but part violates MMC

25.02	
+ 0.25	
Ø25.27	← Actual

as produced:

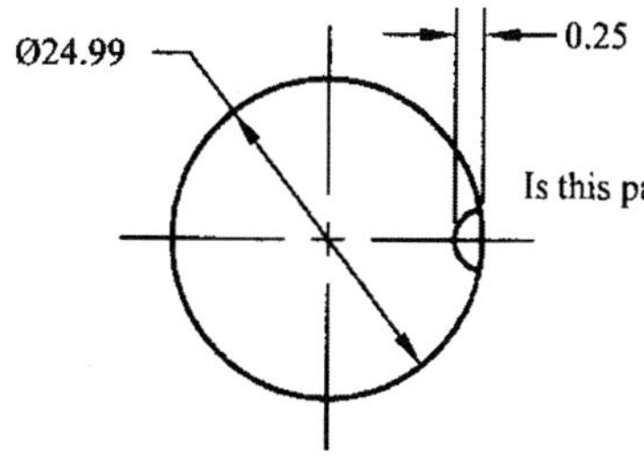

Is this part within tolerance ? No

It violates LMC

24.99	
- 0.25	
Ø24.74	← Actual

13. No. Part is OK. Allowed to taper 0.03 per side, or a diameter of 0.06. It tapers a dia. of 0.04.

14.

as produced:
1st part

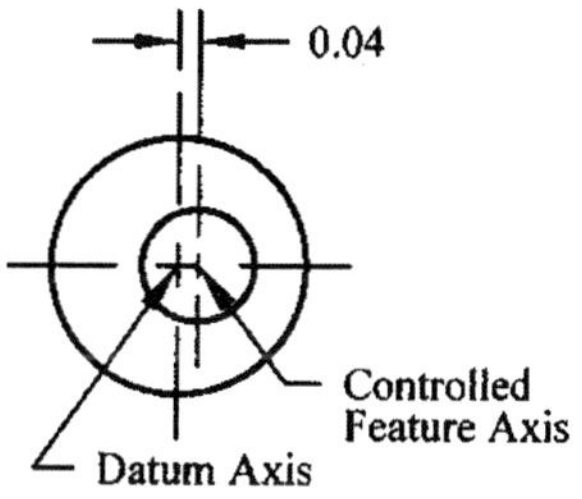

Is the part good?
Why, or why not?

No. Part is only allowed to be off center 0.035.

2nd part

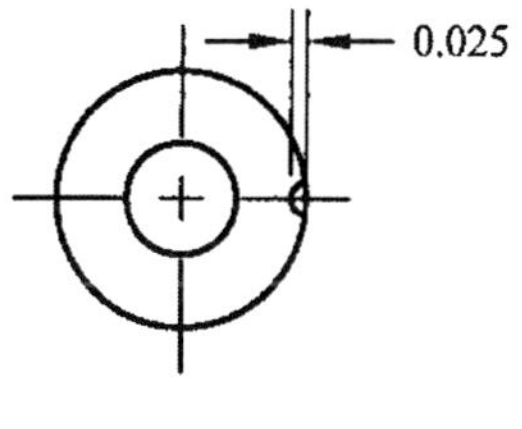

Is the part good?
Why, or why not?

No. It violates the size tolerance.

3rd part

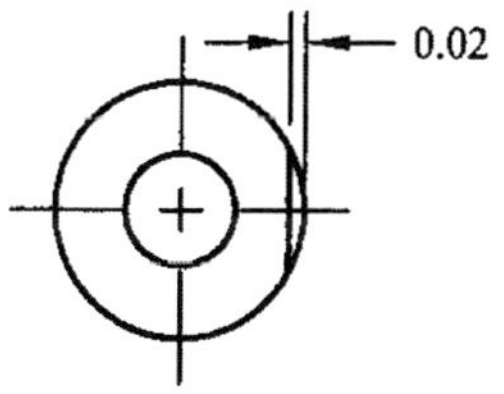

Is the part good?
Why, or why not?

No. It violates circularity control.

15. **Answers page 554**
   1) Size is okay.
   2) Circularity is okay. It doesn't control taper.
   3) Straightness is okay. Don't see any pits, bumps or curves.
   4) Taper is bad. Small diameter is only allowed to taper a diameter of 0.16, but actually tapers a diameter of 0.2.

16. **Answers page 555**
   No. The cylindricity control allows the part to taper 0.01 on each side for a total of 0.02 on the diameter. Circular runout does not control taper, and the size tolerance on the small diameter has not been violated. So, the part does not violate any of the tolerances depicted.

17. **Answers page 556**
   1) Size is okay.
   2) Cylindricity is violated. Allowed to taper a diameter of 0.08 and actually tapers a diameter of 0.11.
   3) Total Runout is violated. It may taper a diameter of 0.14, but does taper a diameter of 0.2.

18. **Answers page 557**
   1st Part: Yes, FIM is 0.07.
   2nd Part: No. Violates size and cylindricity tolerance.
   3rd Part: Yes. FIM is 0.01.

19. **Answers page 558-**Part 12135 – Bushing
   - Outside diameter measures:
     .525 at its largest **OK**
     .519 at its smallest **OK**
     .525-.519= a dia. of .006

   Dia. of .006 actual taper, but cylindricity of .005 allows taper of .005 per side (a diameter of .010)

   - Part outside diameter is round to within .0002. **OK**
   - FIM of I.D. while centered on and rotating about datum axis D is .002 at each cross section. **OK**
   - Inside diameter is:
     .2550 at its largest **OK**
     .2549 at its smallest **OK** **.002 must be off center (eccentric) because inside diameter indicates shape is good.**
   - Part thickness varies between .196 and .195. It is rigid. **OK**
   - Part thickness has a slight uniform bow. It has a centerplane straightness of .003 **No. Exceeds MMC envelope.**

Guide Pin-Answers to page 559

1) .001

2) .002

3) .001

4) A diameter of .010. This is controlled by the size tolerance in the general tolerance note. Circular runout does not control taper.

5) The amount of bonus tolerance for threaded holes allowed by the MMC symbol is not quantifiable.

6) A diameter of .012 (.194 virtual condition – .182 LMC)

7) .010

8) Three sets of two parallel planes .005 apart. These three tolerance zones are oriented at a basic angle of 120° to each other.

PC Board Answers to Page 560

1. .255 x 2.050 x 3.175

2. 2.010

3. 1.990

4. .500 + .025 (half profile tol.) + .005 (half profile tol.) = .530

5. .500 - .025 - .005 = .470

PC Board Answers to Page 561

1. .255 x 2.050 x 3.175

2. 2.010

3. 1.990

4. .500 + .025 (half position tol.) + .005 (half .010 bonus tol. from datum B made at LMC) = .530

5. .500 - .025 - .005 = .470

Answers to page 562

1) .002

2) each other

3) all four – 2 datum holes and the 2 holes indicated Z (through SIM REQT)

4) B

5) perpendicularity

6) total of .010

7) X holes = .281; Z holes = .281

8) C

9) .019; .014; .024

10) pattern of 4 holes; harder to find the central axis

11) optical comparator with dedicated comparator chart; ease and accuracy. This is one of many possible answers. Others might be a Coordinate Measuring Machine (CMM) with scanning probe, Laser Scanners, Optical CMM and many others.

Answers to page 563

1) No – it exceeds the envelope of MMC.

2) 1.963 ± .019

1.963	1.963
+ .019	- .019
1.982	1.944
Yes	

3) No

4) 1.123 ± .024

1.123	1.123
+ .024	- .024
1.147	1.099
	Yes

5) No – at .300 they are allowed .019.

6) Yes – it is out of tolerance (profile is all inside) unless datum shift could bring bump into tolerance zone.

7) Yes – they violate Rule #1 (must receive .295 GO GAGE PIN).

8)

$$\begin{array}{r} .305 \\ +\ .295 \\ \hline .600 \end{array}$$

.600 divided by 2 - .300

$$\begin{array}{r} .300 \\ +\ 1.963 \\ \hline 2.263 \end{array}$$

2.263 ± .019 (half of .038) [2.282 OK]

Workbook Answer to page 564-Crankshaft Assembly

Detail drawing callouts

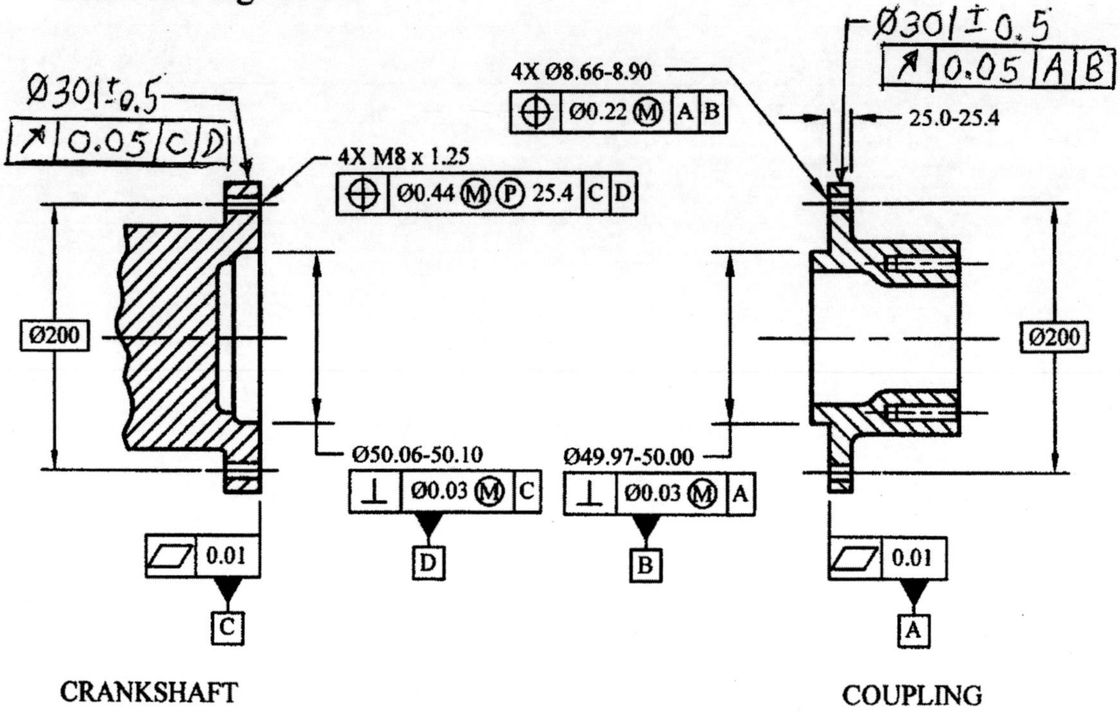

50.06 = MMC
50.00 = MMC
0.06

0.03 for ⊥
2) 0.06

8.66 = MMC
8.00 = MMC
0.66 = Tol to be divided

Workbook Answer to pages 565 and 566

Detail drawing callouts

PART 1

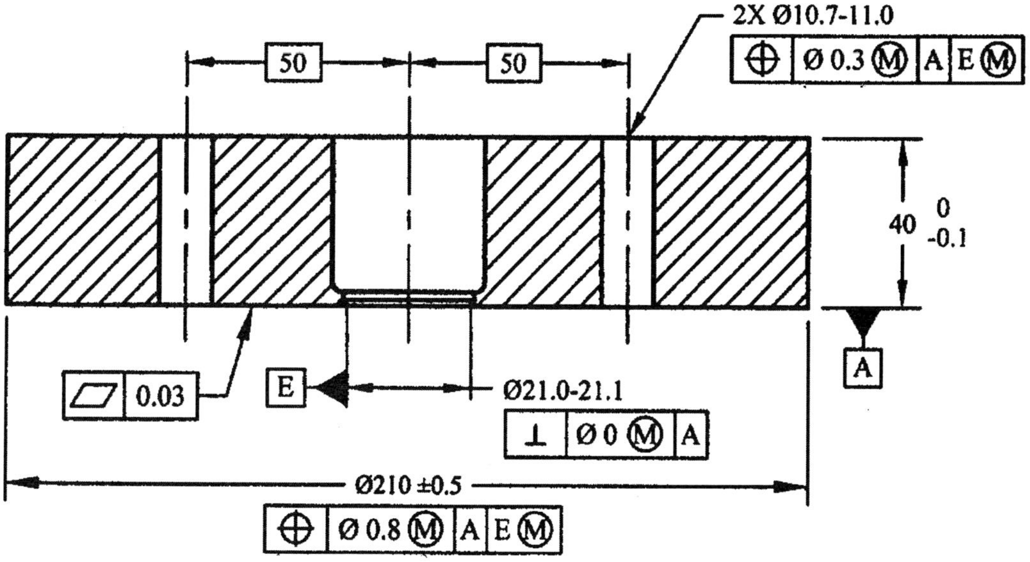

PART 2

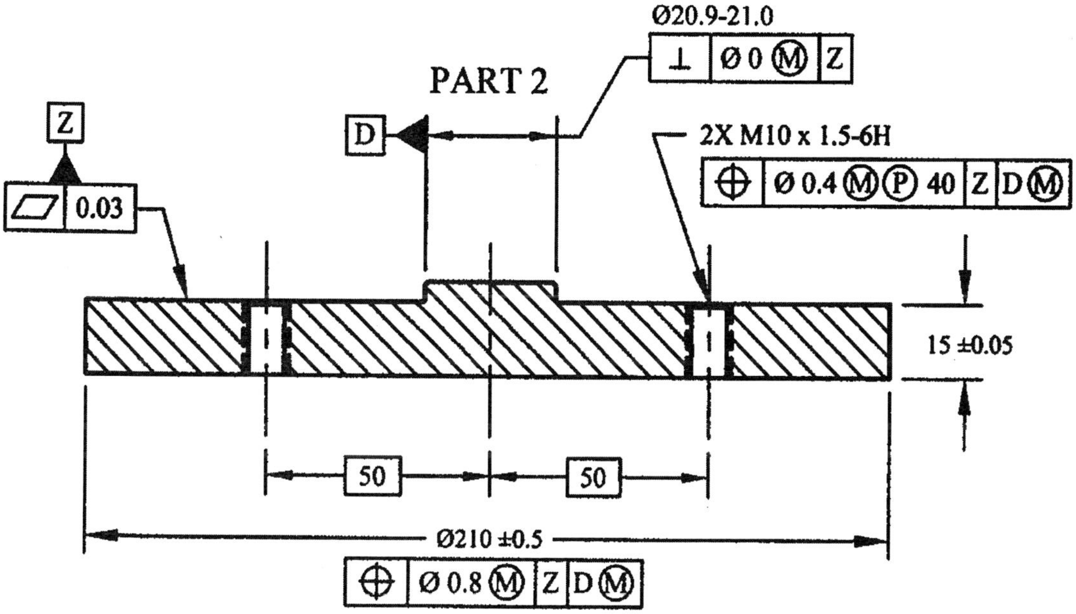

Workbook Answer to pages 567 and 568

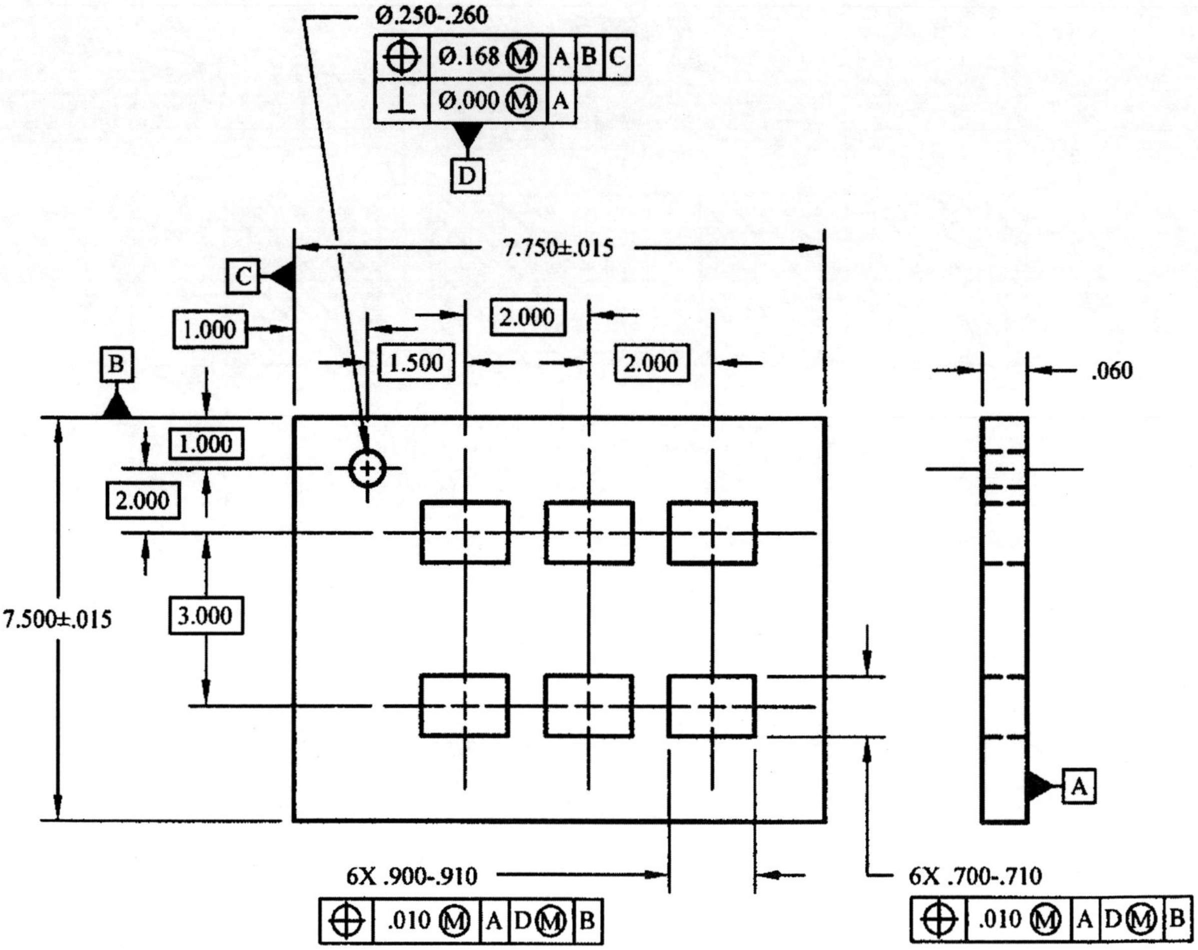

.X = ±.06
.XXX = ±.005
∠ = ±1°

Workbook Answer to page 569

Step 1: Outside Diameter

2.990 = MMC	2.960 = LMC
+ .004 = Geo Tol. at MMC	- .034 = Geo Tol. at LMC
Ø2.994 = Outer Boundary Shaft	2.926 = Inner Boundary Shaft

2.994 = Outer Boundary	2.994 = Outer Boundary
+ 2.926 = Inner Boundary	- 2.926 = Inner Boundary
5.920 = Sum	.068 = Difference

$\frac{5.920}{2}$ = 2.960 Mean

$\frac{.068}{2}$ = .034±Tol.

Divide by 2 to convert to radii:

$\frac{2.960}{2}$ = R1.480

$\frac{.034}{2}$ = ±.017

So,

R1.480±.017.

Step 2: Holes

.570 = MMC	.590 = LMC
- .020 = Geo Tol. at MMC	+ .040 = Geo Tol. at LMC
Ø.550 = Inner Boundary	Ø.630 = Outer Boundary

.630 = Outer Boundary	.630 = Outer Boundary
+ .550 = Inner Boundary	- .550 = Inner Boundary
1.180 = Sum	.080 = Difference

$\frac{1.180}{2}$ = .590 Mean

$\frac{.080}{2}$ = .040±Tol.

Divide by 2 to convert to radii:

$\frac{Ø.590}{2}$ = R.295

$\frac{.040}{2}$ = ±.020

So,

R.295±.020.

Workbook Answer to page 569 continued

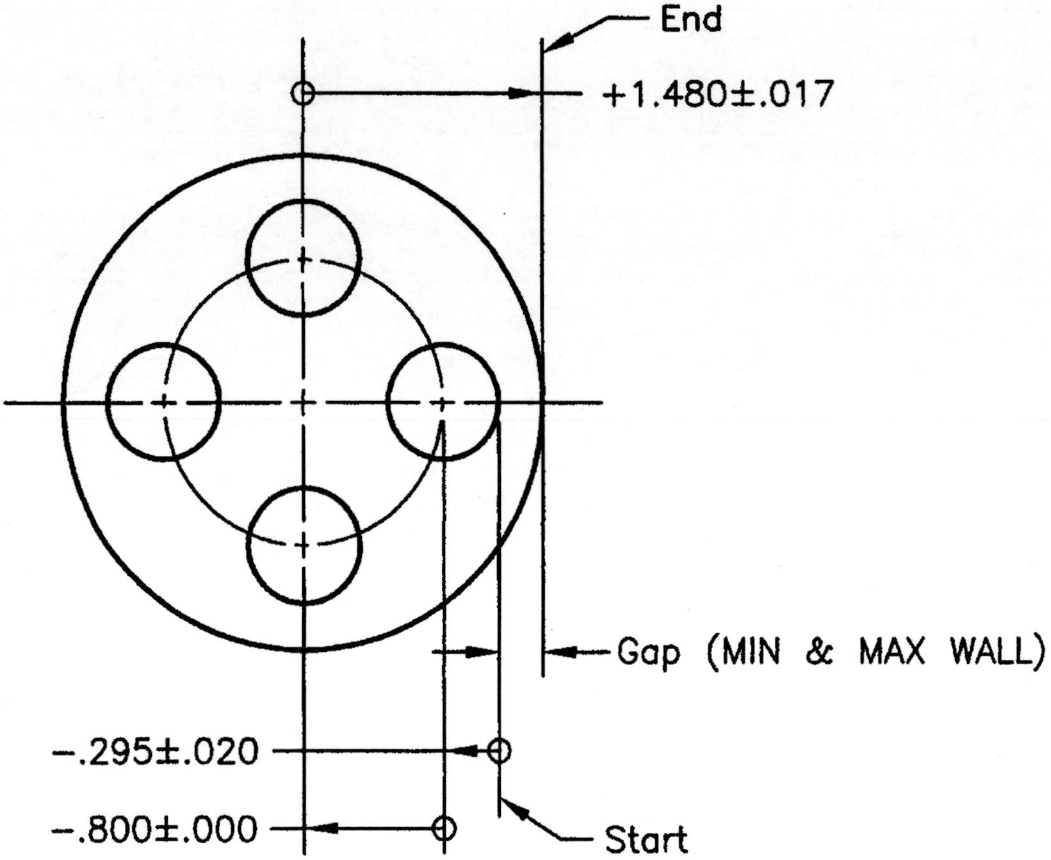

-	+	±Tol	
.295		.020	
.800		.000	
	1.480	.017	
-1.095	+1.480	.037	Totals

$$\begin{array}{r} 1.480 \\ -\ 1.095 \\ \hline .385 \end{array} \qquad \begin{array}{r} .385 \\ -\ .037 \\ \hline .348 \end{array} = \text{MIN WALL} \qquad \begin{array}{r} .385 \\ +\ .037 \\ \hline .422 \end{array} = \text{MAX WALL}$$

Workbook Answer to page 570

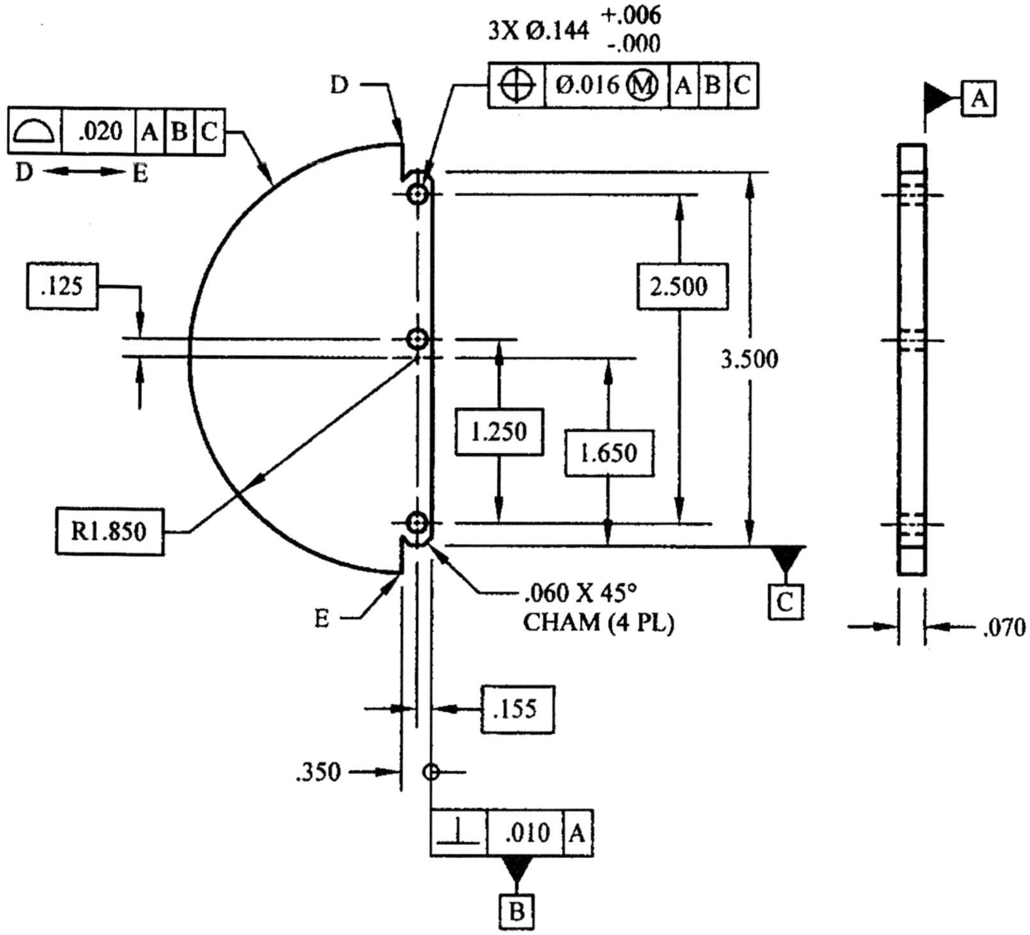

NOTES:

1. TO BE INTERPRETED PER ASME Y14.5
2. ALL GEOMETRIC TOLERANCES TO BE INSPECTED WHILE PART IS RESTRAINED ON PRIMARY DATUM.

UNLESS OTHERWISE SPECIFIED:
.XXX = ±.010
∠ = ±1°

Workbook Answer to page 571

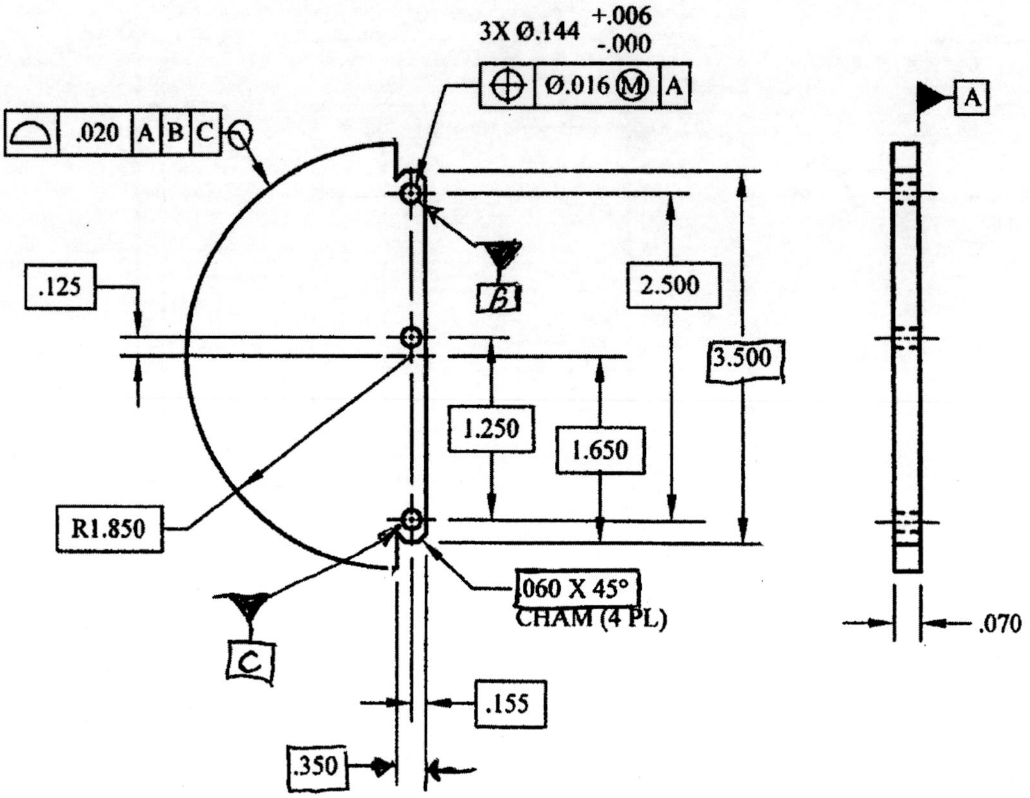

NOTES:

1. TO BE INTERPRETED PER ASME Y14.5
2. ALL GEOMETRIC TOLERANCES TO BE INSPECTED WHILE PART IS RESTRAINED ON PRIMARY DATUM.

UNLESS OTHERWISE SPECIFIED:
.XXX = ±.010
∠ = ±1°

Workbook Answers to pages 572 and 573

1) Yes
2) No 1.963
   + .015
   1.978 MAX allowed
3) Yes
4) No. 1.123
   - .020
   1.103 MIN allowed
5) No. Allowed a diameter of .015 MAX; therefore, part is okay.
6) Not necessarily. With datum feature (pattern) shift the part may be good.
7) Yes.
8) No. The minimum allowed over the pin measurement is calculated as follows:

MIN	**MAX**
1.963 basic dimension	1.963 basic dimension
+ .500 (1/2 of .495 + .505 = 1.000)	+ .500 (1/2 of .495 + .505 = 1.000)
2.463	2.463
- .010 (1/2 of .020)	+ .010 (1/2 of .020)
2.453	2.473
- .005 (1/2 of .010)	+ .005 (1/2 of .010)
2.448 MIN allowed over the pins measurement	2.478 MAX allowed over the pins measurement

-Allowed range 2.448 to 2.478

-Actual measurement of 2.458 is within the allowed range; therefore, it is not indicated that the tolerance has been exceeded.

Workbook Answers to pages 574, 575 and 576

1) How flat must datum feature A be when produced? **.002**

2) For the four holes that become datum pattern D, what are they being positioned to besides datum feature A? **each other**

3) How many holes does the profile control tie the profile to? **all seven**

4) What acts as the origin of measurement for location for the three hole pattern? **the central axis of datum pattern D as generated by its four virtual condition boundaries**

5) What characteristic of geometry does datum A tolerance when used in each feature control frame? **perpendicularity**

6) Is the profile control a total wide zone of .01 or is it ± .01? **a total wide zone of .010**

7) What is the virtual condition of the holes in the three hole pattern? **A diameter of .510**

8) How many gages would be necessary to gage the MMC of all seven holes, the two positional controls and the profile control (if only the outer tolerance zone boundary of the profile control was to be gaged)? **one**

9) If the two outermost .51-.52 holes are produced at a diameter of .514 and their axes a distance apart of 1.126, are they within tolerance? **1.120 + .002 (1/2 of .004) + .002 (1/2 of .004) = 1.124 allowed; therefore, 1.126 is bad**

10) If the surface profile has a bump on the upper left radius that exceeds the basic radius shown by .005, is the surface in violation of the tolerance zone under all circumstances? **with datum feature (pattern) shift, no**

11) If the largest gage pin that can be inserted into the middle .510-.520 hole measures a diameter of .518 and the hole just to its left can receive a gage pin no larger than .512 and a vernier caliper measurement from the outermost points on the gage pin diameters is a distance of 1.080, would that indicate that they exceed their positional tolerance? **.560 + .259 (1/2 of .518) + .256 (1/2 of .512) + .004 (1/2 of .008) + .001 (1/2 of .002) equals 1.080 MAX over the pins meas. allowed**

12) How does the gaging you visualized when answering Question 8 differ from the gaging that would be used for the part on Page 578? **It would be the same.**

Workbook Answers to pages 577 and 578

1) If the holes on the 1.600 diameter bolt circle have been produced at a distance from datum axis B of .820, under what conditions might the holes meet their positional tolerance? **If the holes are produced at a diameter of .590.**

2) If the 2 holes opposite each other on the bolt circle are produced at a distance apart of 1.630 and the holes measure a diameter of .578, do they meet their positional tolerance? **.578 holes get .028 positional tolerance each. So, 1.600 + (1/2 of .028) + (1/2 of .028) = 1.628 allowed; therefore, 1.630 is a NO.**

3) If the center hole is produced at a diameter of .586 and is off center of datum axis B a radius of .017, it is within its positional tolerance? **At a diameter of .586, the positional tolerance is a diameter of .036 (R.018), so .017 is okay. Yes.**

4) If the part thickness measures .626 with a vernier caliper and curves like a banana .006 in the right-side view, does it meet its size requirements? **.626 + .006 = .632 which violates MMC envelope, so the answer is no.**

5) If datum feature B is produced at a diameter of 2.976 and is perfectly perpendicular to datum A, how much pattern shift would be allowed for the 5-hole pattern? **2.994 – 2.976 = a diameter of .018 or a radius of .009.**

6) If the outside diameter is produced at a size that will fit into a hole that is 2.980 and seat on datum feature A with three high points of contact, how much perpendicularity tolerance is the O.D. allowed? **2.994 – 2.980 = a diameter of .014**

7) If none of the allowed perpendicularity tolerance in question #6 was used by the O.D., what could the tolerance be used for instead? **Datum Feature (or Pattern) shift of the 5 holes.**

8) If gage pins were inserted into two opposite holes on the 1.600 bolt circle and a measurement taken over the pins was 2.230, could the holes be within their positional tolerance? **Yes.** Under what circumstances? **If holes are made at a diameter of .590, then 1.600 + (1/2 of .590) + (1/2 of .590) + (1/2 of .040) + (1/2 of .040) = 2.230 allowed.**

9) If two holes on the bolt circle were produced at a diameter of .580, what would their maximum allowed axial separation be? **Positional tolerance at .580 = .030, so 1.600 + (1/2 of .030) + (1/2 of .030) = 1.630**

10) What is the maximum amount of pattern shift for the five-hole pattern that may ever occur on this part? **2.994 – 2.960 = a diameter of .034 or R.017**

Workbook Answers to page 579

FIXED FASTENER SITUATION

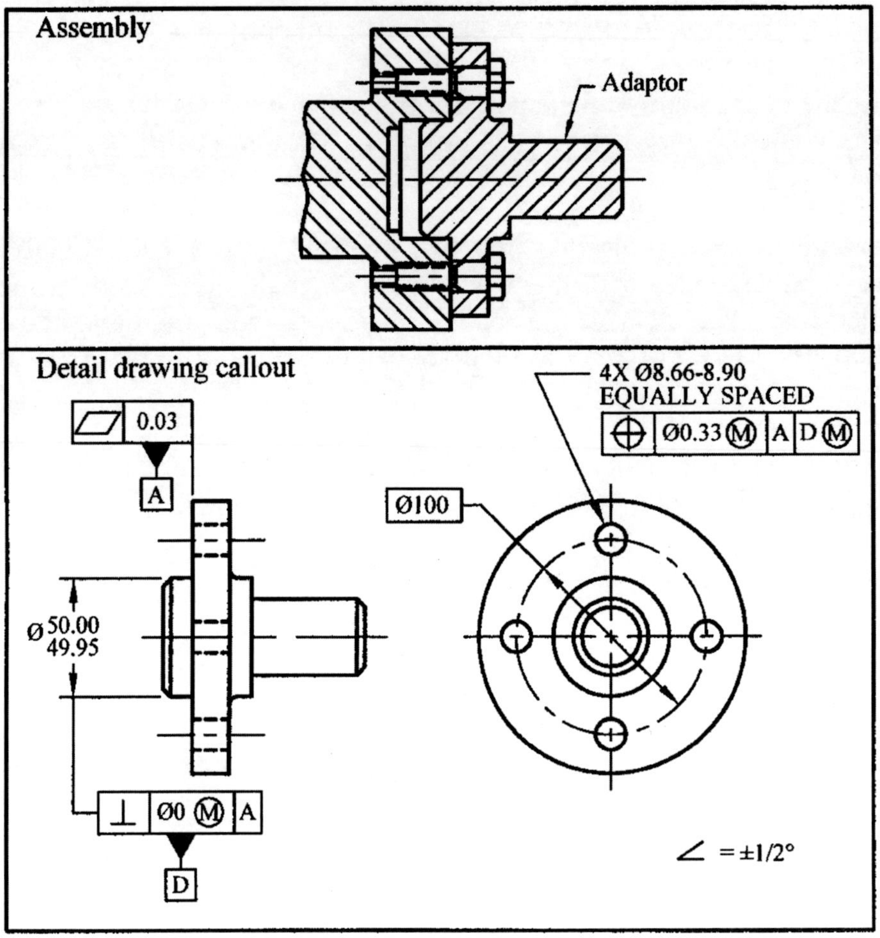

1. Locate the four-hole pattern. To calculate the positional tolerance, use an 8mm screw.

$$\begin{array}{rl} 8.66 & = \text{MMC Hole} \\ -\ 8.00 & = \text{MMC Fastener} \\ \hline 0.66 & \end{array}$$

$$2\overline{)0.66} \quad 0.33$$

Extra information: The bore that mates with the 49.95-50.00 diameter has a virtual condition of 50.00.

Sequence reminder: 1) flatness 2) perpendicularity 3) position

$$\begin{array}{r} 50 \\ -\ 50 \\ \hline 0 \end{array}$$

The positional tolerance of 0.66 may be divided unequally between this part and its mating part that has the threaded holes. Perhaps, one would consider giving a greater portion of this tolerance to the part that is most difficult to make. Many would consider threaded holes harder to manufacture than clearance holes, so a division of the tolerance that gives more than half of the 0.66 to the threaded holes and less to these clearance holes would be justified.

Workbook Answers to pages 580 and 581

Assembly

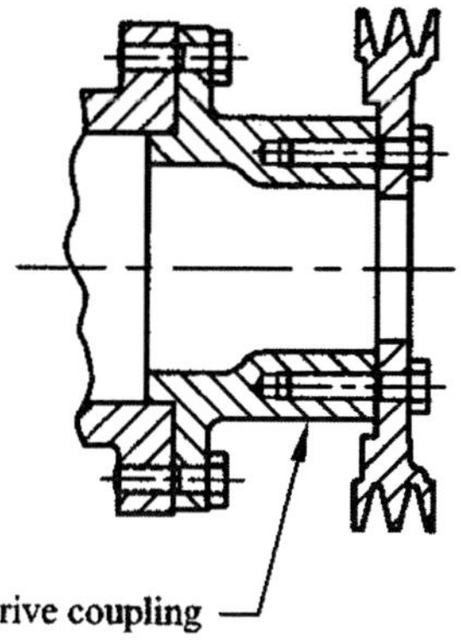

Detail part and gage for checking position of M8 x 1.25 tapped holes

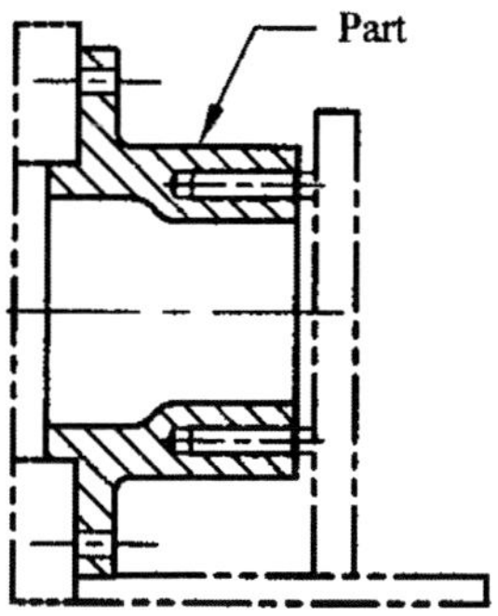

Detail drawing callouts

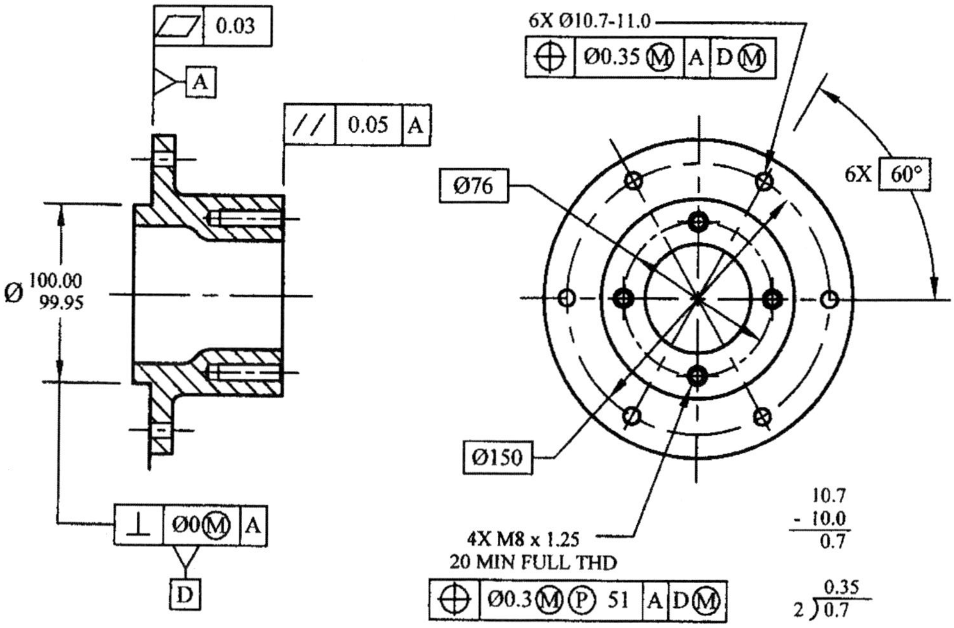

Workbook Answers to pages 582 and 583

Part 1

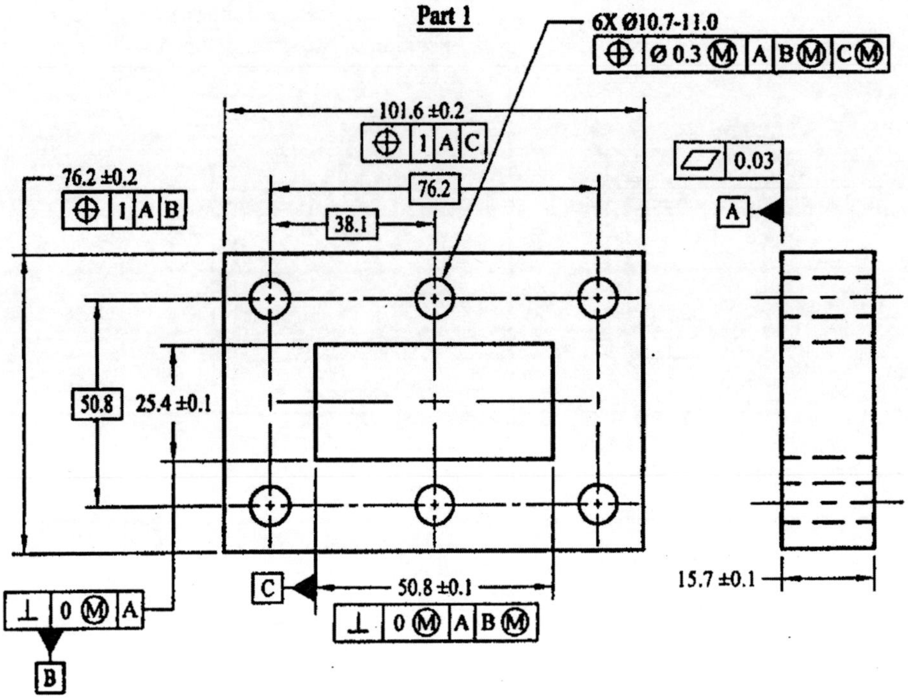

UNLESS OTHERWISE SPECIFIED ALL ANGLES ARE ±1°

Part 2

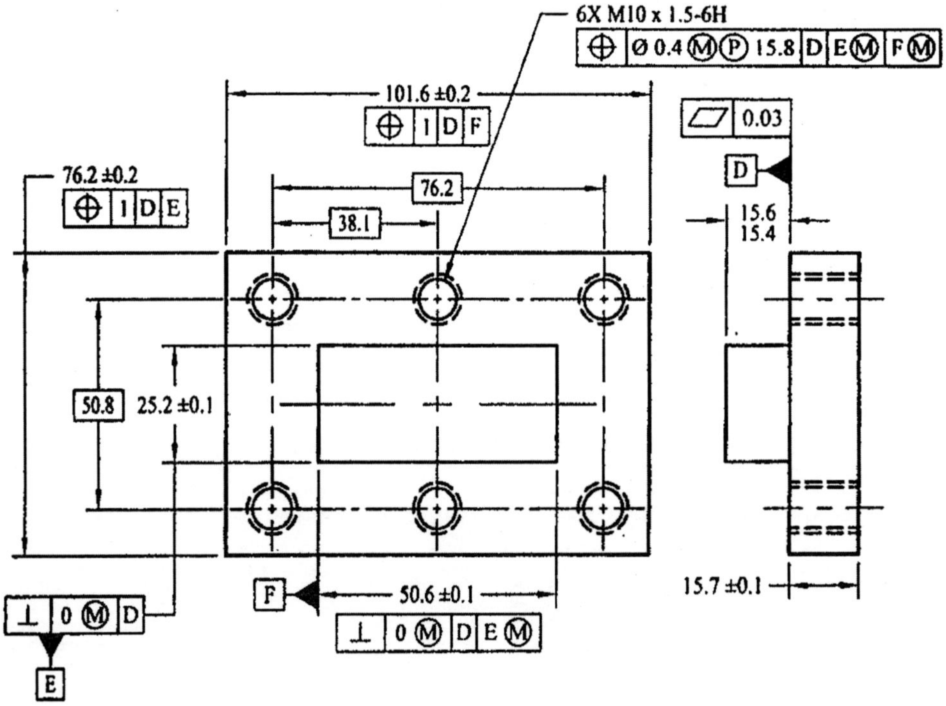

UNLESS OTHERWISE SPECIFIED ALL ANGLES ARE ±1°

Workbook Answer to Exercise on page 584

Step 1: Outside Diameter

75.1 = LMC
- 0.9 = Geo Tol. at LMC
74.2 = Inner Boundary Shaft

Step 2: Holes

15 = LMC
+1 = Geo Tol. at LMC
Ø16 = Outer Boundary

Step 3:

74.2=Inner Boundary
-40.0=Bolt Circle Diameter
34.2= Total Material with hole

34.2=Total Material with hole
-16.0=Outer Boundary of Hole
18.2=Total Material on Both Sides of Hole

18.2 divided by 2= **9.1 MINIMUM WALL THICKNESS BETWEEN ONE HOLE AND OUTSIDE DIAMETER OF THE PART**

Workbook Answers to page 585

Step 1:

.260 = LMC Hole
+ .050 = Geo. Tol. at LMC

Ø.310 = Outer Boundary Hole

+ .130 = TTol. of D at LMC

.440
+ .130 = Pattern Shift (D μ)

Ø.570 = Outer Boundary Hole with all Factors

.240 = MMC Hole
- .030 = Geo. Tol. at MMC

Ø.210 = Inner Boundary Hole

- .130 = TTol. of D at LMC

.080
- .130

Ø -.050 = Inner Boundary Hole with all Factors

+.570
+ -.050
.520

$\frac{.520}{2}$ = .260 = Mean Dimension

and

+.570
- -.050
.620

$\frac{.620}{2}$ = .310 = ± Tol.

So,

Ø.260±.310
R.130±.155

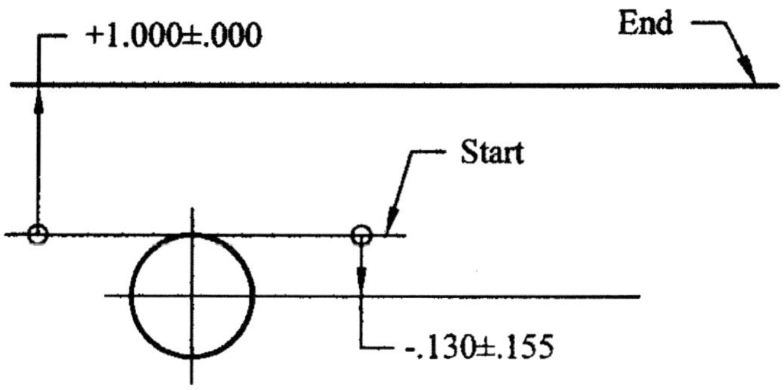

1.000 + .130	.870 + .155	.870 - .155
.870	1.025= MAX WALL to datum plane & datum feature C	.715= MIN WALL

1/2 degree=.008 (over this part's maximum thickness)

6.100±.020=6.120-6.080=.040 flatness tolerance for datum feature C.

.715-.048=.667 MIN WALL to lowest point on the surface of datum feature C.

OR:

.260=LMC Hole
+.050=Geo. Tol. at LMC
.310= Outer Boundary Hole
+.130=D's Position Tol.
.440= Outer Boundary of Hole Moving with D to C
+.130=D's Datum Feature Shift Due to MMB Modifier
.570= Outer Boundary of Hole with D's Position Tol. and Datum Feature (Pattern) Shift Away from D

.570 divided by 2=**.285**=Radius of all Factors Shown Above

1.000=Basic Dimension from C (implied) minus .285=**.715** Min. Wall to Datum Plane C

.715 minus .040 (Datum feature C's flatness tolerance given by overall size dimension=**.675**

.675 minus .008 (one-half degree over .900) =**.667 MIN WALL THICKNESS**

Workbook Answers to page 586

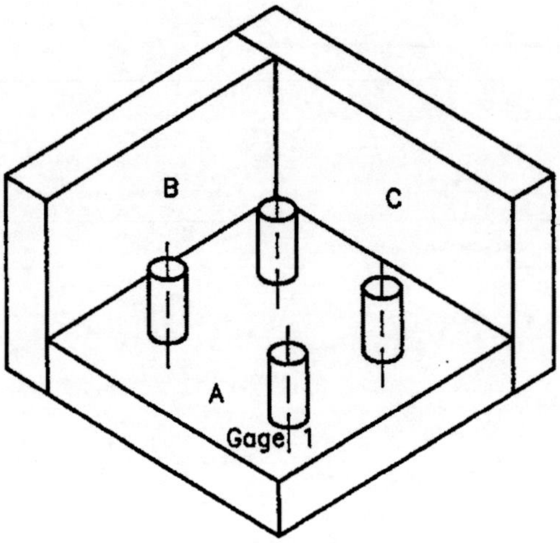

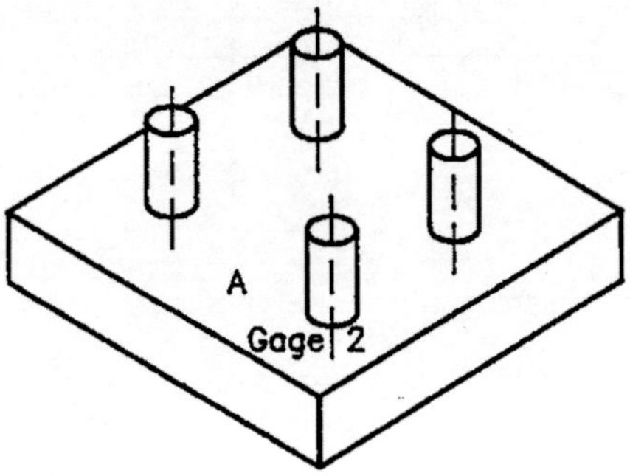

GAGE 1 – Datum to pattern
All four pins are located by the same basic dimensions (from the datum feature simulators) as used on the part drawing.
Pin sizes are ⌀.355 $^{+.XXX}_{-.000}$

with | ⌖ | ⌀.000 Ⓜ | A | B | C |

GAGE 2 – Feature to feature
pin axis to pin axis basic dimensions are the same as gage 1 (and part drawing).
Pin sizes are ⌀.360 $^{+.XXX}_{-.000}$

with | ⌖ | ⌀ .xxx Ⓜ | A | | |
| | ⌀.000 Ⓜ | A |

Workbook Answers to page 587 and page 588

GAGE 1	1a) **Yes**	1b) **.870**
GAGE 2	2a) **Yes**	2b) **.900**
GAGE 3	3a) **.900**	3b) **.500**

Section 6 Workbook Answers pages 589-596

Answer to Workbook page 589

Shafts

30.0 = LMC Shaft
- 1.5 = Geo. Tol. at LMC
28.5 = Inner Boundary of Shaft

Holes

20.00 = LMC Hole
+ 0.25 = Geo. Tol. at LMC
20.25 = Outer Boundary of Hole

28.50 = Inner Boundary of Shaft
- 20.25 = Outer Boundary of Hole
8.25 = Material with Hole

8.25 divided by 2 = 4.125 Minimum Wall Thickness

Answer to Workbook page 590

Minimum wall calculation answer:

LMC	4.990
-Geo. Tol. at LMC	- .040
Minimum Inner Shaft Boundary	4.950 = Inner Boundary of the O.D.

Because the 4 holes and the 4.990 - 5.000 O.D. are located to the same datums in the same order of precedence, they are considered one pattern, a Simultaneous Requirement; therefore, as D departs from MMC, they may shift--but all must shift in the same direction. So, if they all shift in the same direction, their relationship to one another does not change. Therefore, the thickness between the two is not further endangered.

4.950 = Inner Boundary of the Shaft

.375 = LMC
+.035 = Tol. at LMC
.410 = Outer Boundary of the (Hole)

1.600 = Bolt Square
x 1.4142 = Square Root of 2
2.26272 = Diagonal of Square (Bolt Circle)

4.950 = Shaft O.D. Inner Boundary
-2.263 = Diagonal of Square (Bolt Circle)

2.687 = Inner Boundary of the Shaft without the Bolt Circle
- .410 = Outer Boundary of the Hole
2.277 = Total Material

2.277 divided by 2 = 1.1385 Minimum Wall Thickness

Answer to Workbook page 591

4.990 = LMC of O. D.
- .040 = Geo. Tol. at LMC
4.950 = Resultant Condition of O.D.

.375 = LMC of Holes
+.035 = Geo. Tol. at LMC
.410 = Resultant Condition of Holes

1.600 = Bolt Hole Square
x 1.4142 = Square Root of 2
2.263 = Diagonal of Square (Diameter of Bolt Hole Circle)

4.950 = Resultant Condition of O.D.
- 2.263 = Diameter of Bolt Hole Circle
2.687 = Resultant Condition of Shaft w/o Bolt Hole Circle

2.687 = Resultant Condition of Shaft w/o Bolt Hole Circle
- .410 = Resultant Condition of Holes
2.277
- .400 = Movement of D at LMC Allowed to O. D.
1.877

1.877
- .400 = Displacement (Datum feature shift Allowed)
1.477 = Total Material from D at Its As-Produced Location

1.477 divided by 2 = **.7385 minimum wall thickness** of 4-hole pattern to O. D. of the part (datum feature B)

Logic of adding displacement to 4 holes as a pattern

1. Datum feature D may be out of position a diameter of .400 to the outside diameter of the part if it is produced at its LMC of 1.000.

2. As datum regular feature of size D departs from its MMB of .600 to its LMC of 1.000, the pattern controlled to it may shift with the datum feature axis from the datum axis a like amount (one for one). This additional displacement is allowed due to the MMB symbol next to datum feature D in the positional tolerance for the 4 hole pattern.

Answer to Workbook page 592

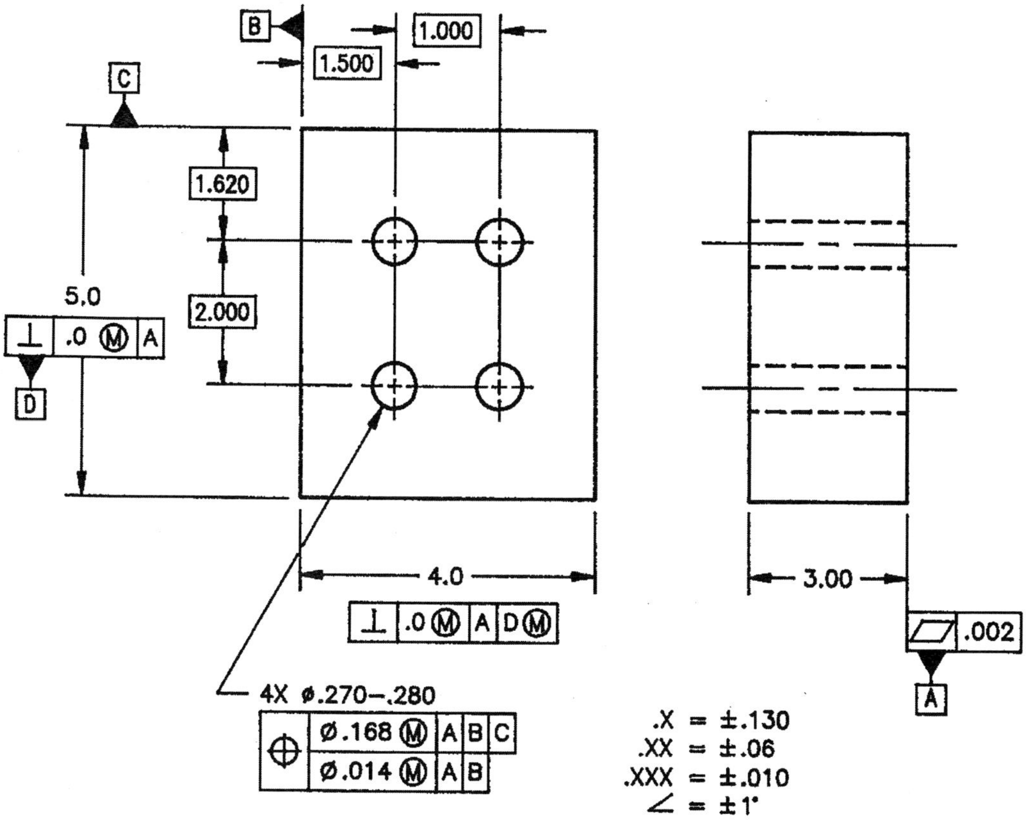

① ±.06 = .120

②
```
   .120
X  1.4
  ø.168
```

③
```
   .270
−  .256
  ø.014
```

Answer to Workbook page 593

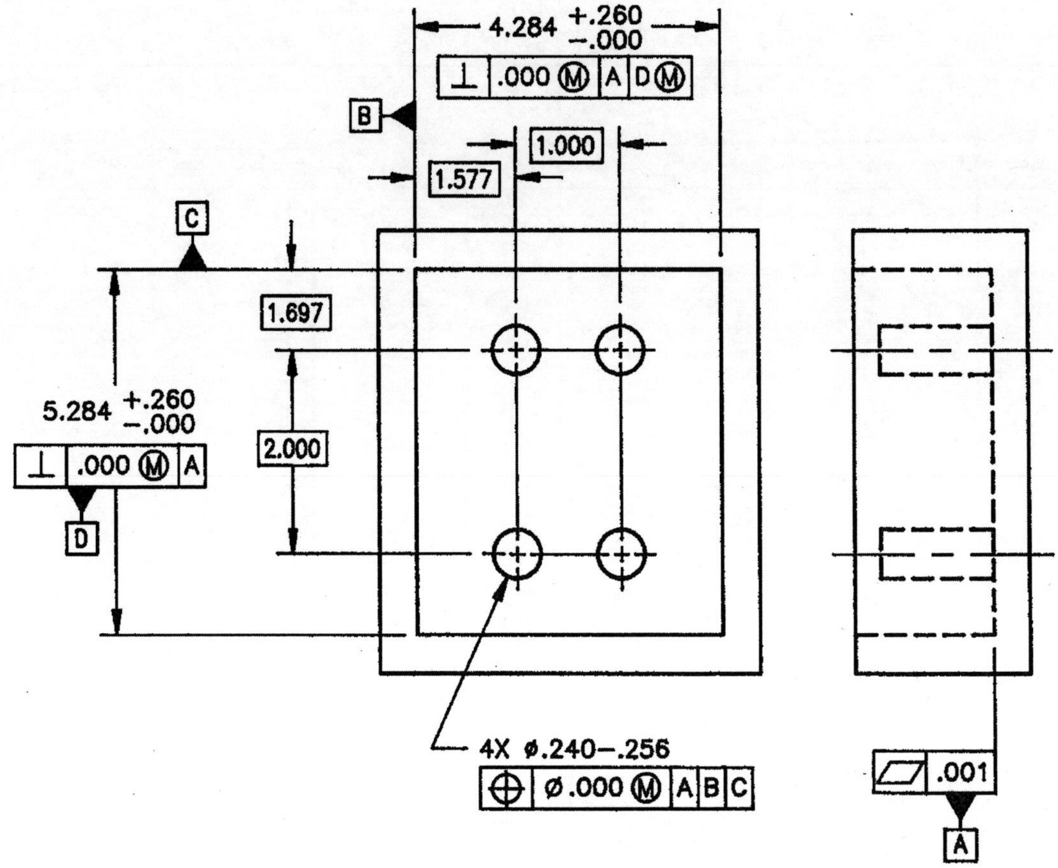

---

① .168 − .014 = .154

② .077 (2) .154)

③ 1.500 + .077 = **1.577**

④ 1.620 + .077 = **1.697**

⑤ 5.000 + .130 = 5.130

⑥ 5.130 + .154 = 5.284

⑦ 4.130 + .154 = 4.284

⑧ .270 = MMC
− .014 = Total MMC
.256 = Virtual Cond.

Answer to Workbook page 594-596

a. .070=LMC of Nail Head
- .007=Position Tol. On Core Board
.063=Diameter of Area for Pogo Pin to Hit

b. .507=LMC of Guide Holes
-.494=LMC of Guide Pins
.013=Diameter of Clearance

c. .050=Position Tol. On Pogo Pins
+.013=Slide of Pogo Pins on Areas to Contact due to Clearance bet. Guide holes & pins
.063=Total Movement of Pogo Pins

d. No. .063
-.063
.000=Possible Contact

SECTION 7-Answers to workbook pages 597-611

Answers to pages 597-601 are shown in questions 1-19.

1. Please complete the following feature control frame to indicate that the datum feature simulator for datum feature C translates (moves).

   21X Ø 4.0-4.3

⊕	Ø 0 Ⓜ	A	B Ⓜ	C Ⓜ ▷

2. Complete the following feature control frame to indicate a unilateral profile of a surface control that is 0.8 all inside of the basic profile.

⌓	0.8 Ⓤ 0	A	B Ⓜ	C Ⓜ

3. Eliminate the envelope of perfect form at maximum material condition by using the appropriate symbol after the size limits of the following hole.

   Ø 10.5-10.7 Ⓘ

4. Use the correct symbology to make the following two diameters treated as though they are one diameter.

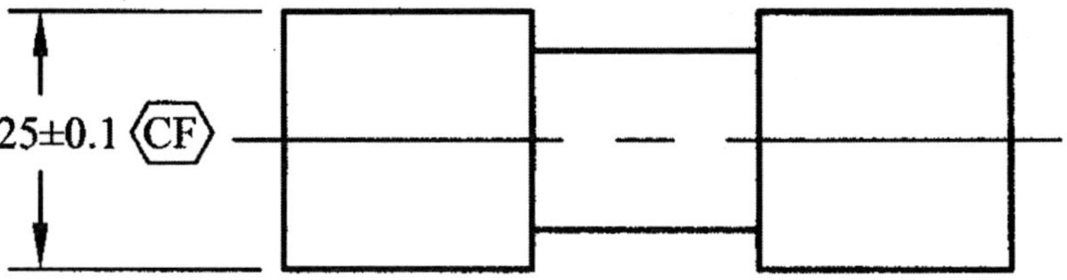

5. When the symbol Ⓜ is used after a datum feature of size referenced in a feature control frame, it stands for: (Circle correct answer)
   a. Minimum Material Condition
   b. Maximum Material Condition
   c. Maximum Material Boundary
   d. Minimum Material Boundary

6. When the symbol Ⓛ used after a geometric tolerance in a feature control frame, it stands for: (Circle correct answer)
   a. Least Material Boundary
   b. Least Material Condition
   c. Least Squares Center
   d. Least Boundary Condition

7. Put a symbol in the blank area on the elbow of the leader line from the feature control frame that means ALL OVER, APPLIES IN ALL VIEWS SIMULTANEOUSLY.

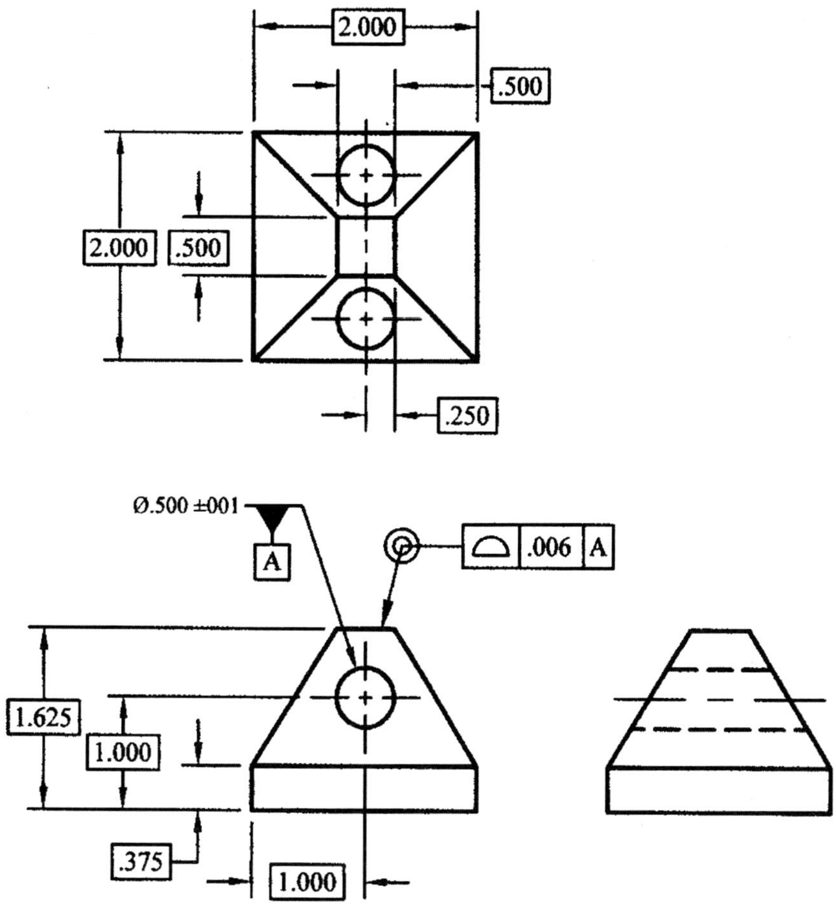

8. When no material condition symbol is used inside a feature control frame after any datum features of size referenced, what is the abbreviation for the symbol that is implied? (Circle correct answer)
   a. MMC
   b. RMB
   c. LMC
   d. MMB
   e. RFS
   f. LMB

9. What does each of the abbreviations in question 8 stand for?
   a. MMC stands for **Maximum Material Condition.**
   b. RMB stands for **Regardless of Material Boundary.**
   c. LMC stands for **Least Material Condition.**
   d. MMB stands for **Maximum Material Boundary.**

e. RFS stands for **Regardless of Feature Size.**

f. LMB stands for **Least Material Boundary.**

10. What does the symbol that includes the SF letters in the following drawing stand for? **Spotface**

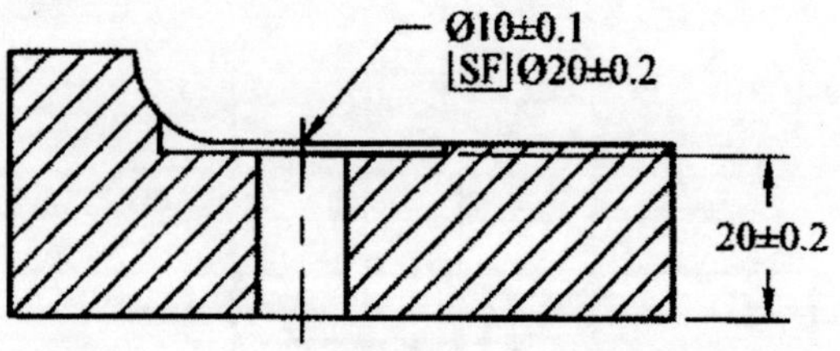

11. According to the rules in ASME Y14.5-2009 and 2018, may the Ⓢ symbol be used in a feature control frame? **No. It is implied, so it may not be specified.**

12. Label each of the envelopes to which arrows point in the following illustrations as either the "actual mating envelope" or the "actual minimum material envelope".

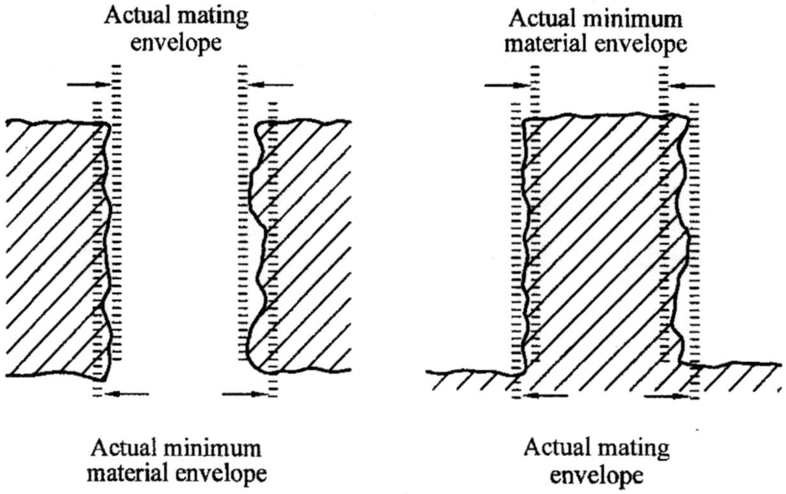

INTERNAL FEATURE

EXTERNAL FEATURE

13. What geometric characteristic symbol (if placed below the 25±0.5) would be capable of creating the following tolerance zone for the part shown? (Circle the correct answer below the illustrations)

Part Specification

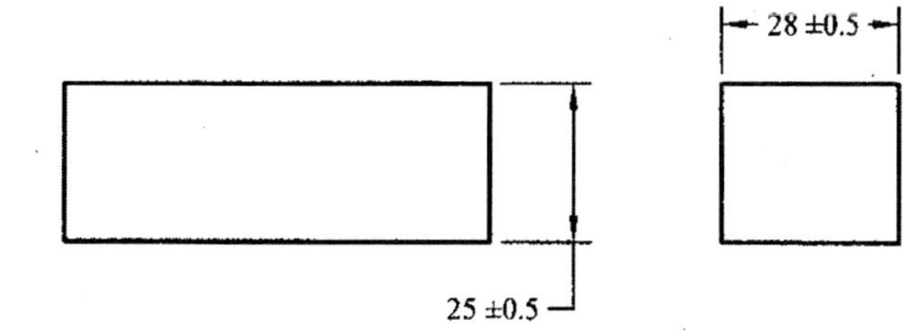

Tolerance Zone

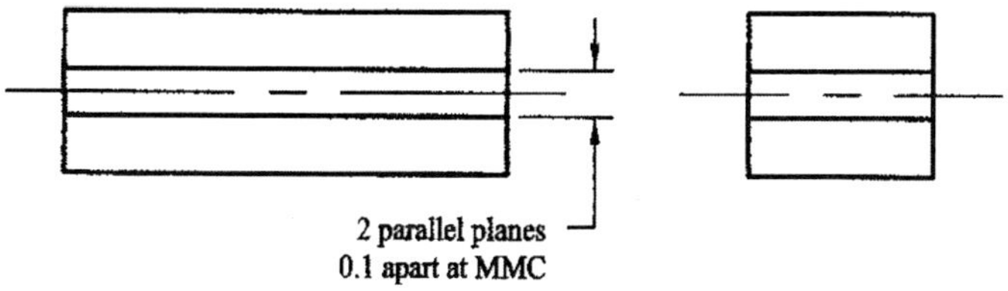

a. 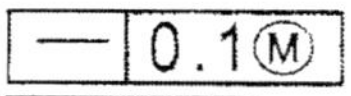

b. 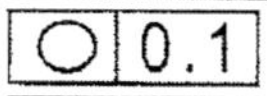

c. 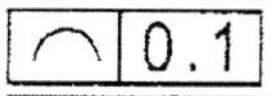

d. 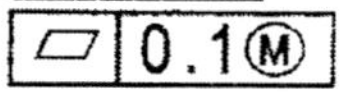

14. An "Actual Mating Envelope" is a similar perfect feature(s) counterpart of smallest size that can be contracted about an external feature(s) or of largest size that can be expanded within an internal feature(s) so that it coincides with the surface(s) at the highest points. What is the difference between an "Unrelated Actual Mating Envelope" and a "Related Actual Mating Envelope"? (Circle one)
   a. Unrelated is not tied to the feature control frame, but related is.
   b. Related is oriented or located to a datum(s), while unrelated is not.
   c. Unrelated is not a "Simultaneous Requirement" with other feature patterns, but Related is related to other feature patterns.
   d. Related is a combined control using one feature control frame stacked on another.

15. Which geometric characteristic symbol that is categorized as "Orientation" may be used to tolerance any angle no matter how many degrees it is specified as?
   a. Angularity
   b. Perpendicularity
   c. Parallelism
   d. Profile of a Surface

16. Where axes, center planes or surfaces are shown coincident on a drawing and geometric tolerances establish the relationship between the features, what applies?
   a. a plus and minus tolerance
   b. a zero basic dimension

c. a general tolerance note
d. a reference dimension

17. Where a coordinate system is shown on the drawing, unless otherwise specified, it shall be:
   a. left handed
   b. labeled in the negative direction
   c. neither labeled in the positive or negative directions
   d. right handed

18. Per **ASME Y14.5-2009**, unless otherwise specified, a datum feature simulator is:
   a. a true geometric counterpart
   b. physical
   c. theoretical
   d. rails, machine table, surface planes, collet chucks and parallels used to fixture parts.

However, in Y14.43 a datum feature simulator has always been defined as a physical representation of a datum feature such as rails, a machine table, surface plates, collet chucks and parallels used to fixture or functionally gage parts.

What was called a datum feature simulator that was perfect and theoretical in ASME Y14.5-2009, is called in ASME Y14.5-2018 a true geometric counterpart (TGC) as it was in ASME Y14.5-1994. One should be aware of the year of the Y14.5 revision they are using and what the rules and terms are in that revision, because some things change in standards when they are revised.

19. Match the following symbols with what they represent by listing the correct letter of the definition after each symbol.
   a. Continuous Feature of Size (treats multiple features of size as one)
   b. Translation symbol (datum feature simulator translates/moves)
   c. All Over (control applies in all views-commonly used with profile of a surface)
   d. Unequally Disposed or Unilateral Tolerance
   e. Independency (size tolerance does not control form)

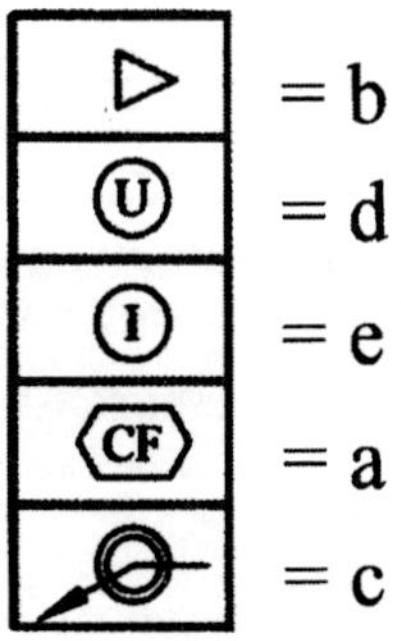

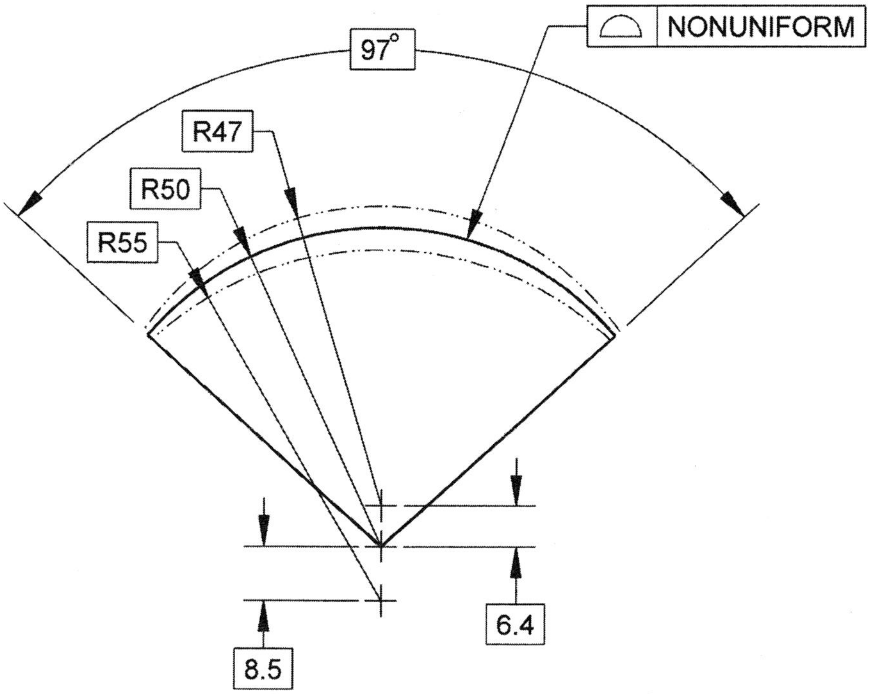

Answer to workbook page 602

20) The illustration above is of a NONUNIFORM profile of a surface control. The surface being toleranced by this control must reside within the tolerance zone depicted by the phantom lines. The purpose of such a tolerance zone can be to: (Circle all correct answers)

a. smooth the transition between profile segments

b. allow greater variations in the surface form at the ends of the zone

c. allow abrupt surface variations at the ends of the profile zone

d. allow smaller surface variations at the ends of the profile zone

e. allow larger surface variations in form in the middle of the zone

Answers to workbook page 603

21. The following illustration compares Composite Profile Tolerancing (one profile symbol) with Two Single Segment Profile Tolerancing (two profile symbols).

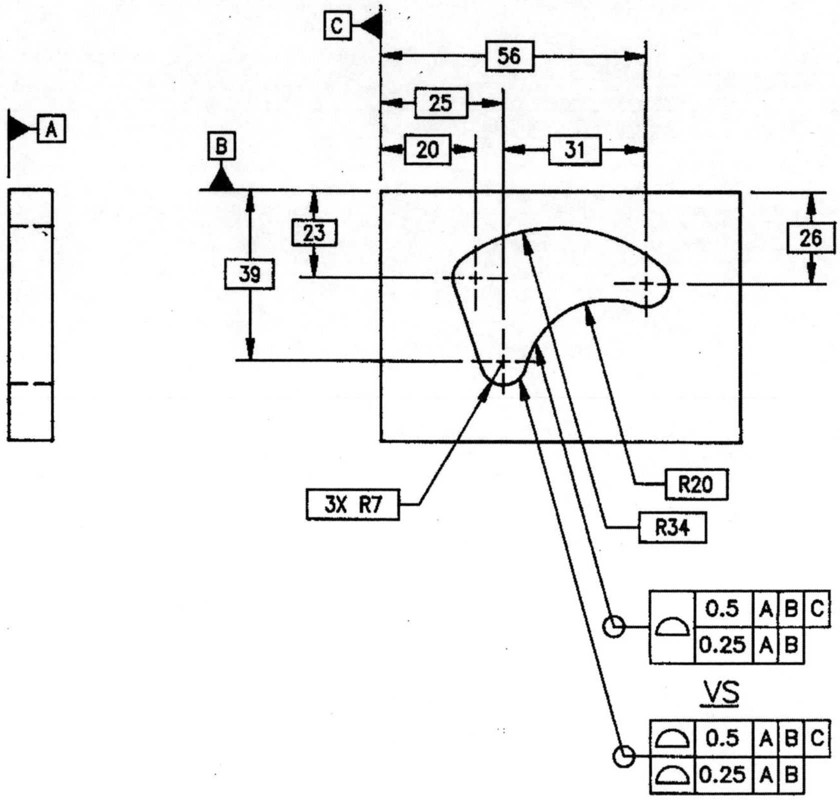

In the composite profile control:

a. What is the tolerance on the size and form of the hole? **0.25**
b. What is the tolerance on the perpendicularity of the hole to datum plane A? **0.25**
c. What is the tolerance on the angular orientation of the hole to datum plane B? **0.25**
d. What is the tolerance on the location of the hole to datum plane B? **0.5**

In the two single segment profile control:

e. What is the tolerance on the size and form of the hole? **0.25**
f. What is the tolerance on the perpendicularity of the hole to datum plane A? **0.25**
g. What is the tolerance on the angular orientation of the hole to datum plane B? **0.25**
h. What is the tolerance on the location of the hole to datum plane B? **0.25**

Answers to workbook page 604

22. The following illustration a composite profile control with another profile control that applies INDIVIDUALLY to each hole.

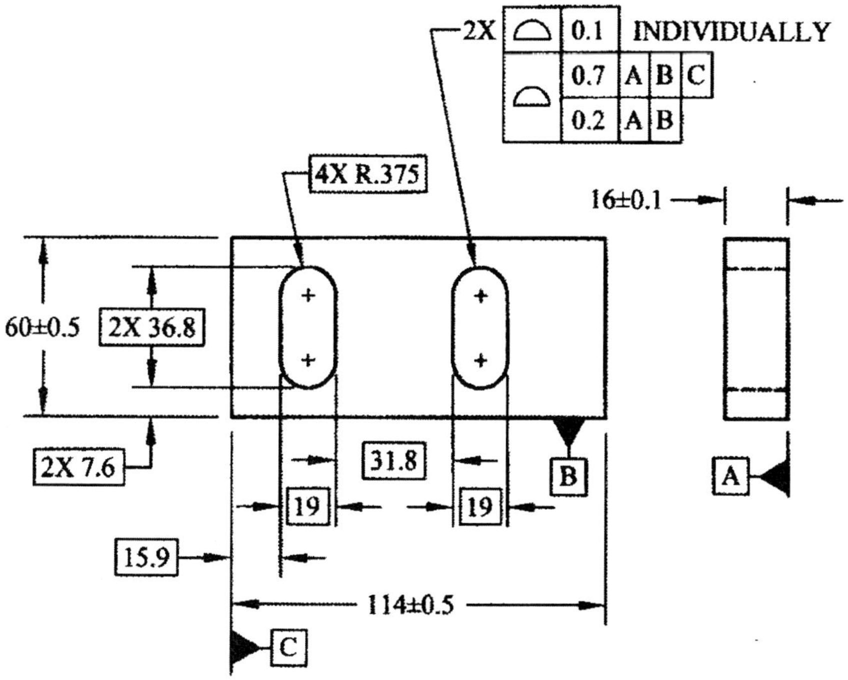

a. What is the tolerance on the size and form of the holes? **0.1**
b. What is the tolerance on the location of the holes to datum B? **0.7**
c. What is the tolerance on the location of the holes to datum C? **0.7**
d. What is the tolerance on the location of the holes to each other? **0.2**
e. What is the tolerance on the perpendicularity of the holes to datum plane A? **0.2**
f. What is the tolerance on the angular orientation (rotation) of the 2-hole pattern to secondary datum plane B? **0.2**
g. What is the tolerance on the angular orientation (rotation) of the 2-hole pattern to tertiary datum plane C? **0.2** **(If you cannot rotate more than 0.2 to B, you cannot rotate more than 0.2 to C, since datum planes B and C are perpendicular to each other.)**

Answers to workbook page 605

23. The following illustration shows a conical feature used as a datum feature.

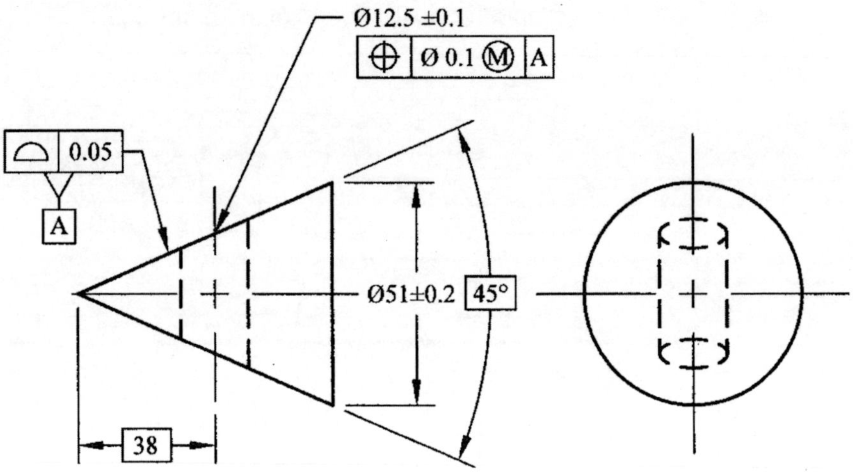

a. How many spatial degrees of freedom are controlled by referencing datum feature A in the position control on the hole? **5**

b. Does datum feature A generate:
   1. an axis only
   2. an axis and a plane (from the point at the apex of the cone)
   3. an axis and a plane from the right edge in the view on the left
   4. a plane only

24. The following illustration shows a conical datum feature. In the position feature control frame the spatial degrees of freedom that each datum reference controls are listed in brackets after the datum reference.

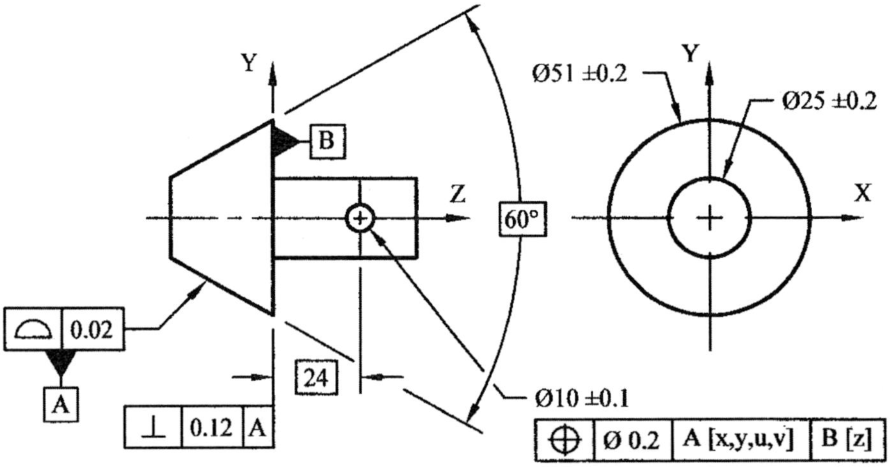

a. What controls the rotational spatial degree of freedom known as w? **Nothing**

b. If datum feature B was referenced as primary instead of secondary, would it be necessary to specify the bracketed z after B in the feature control frame? **No**

c. If datum feature B was not referenced in the position control, how would the dimensions on the part drawing change? **The 24mm basic dimension would go away and be replaced by a basic dimension from the apex of the cone.**

Answers to workbook page 606

25. Detail Drawing with Degrees of Spatial Freedom Indicated by Showing X, Y and Z Axes on the Drawing.

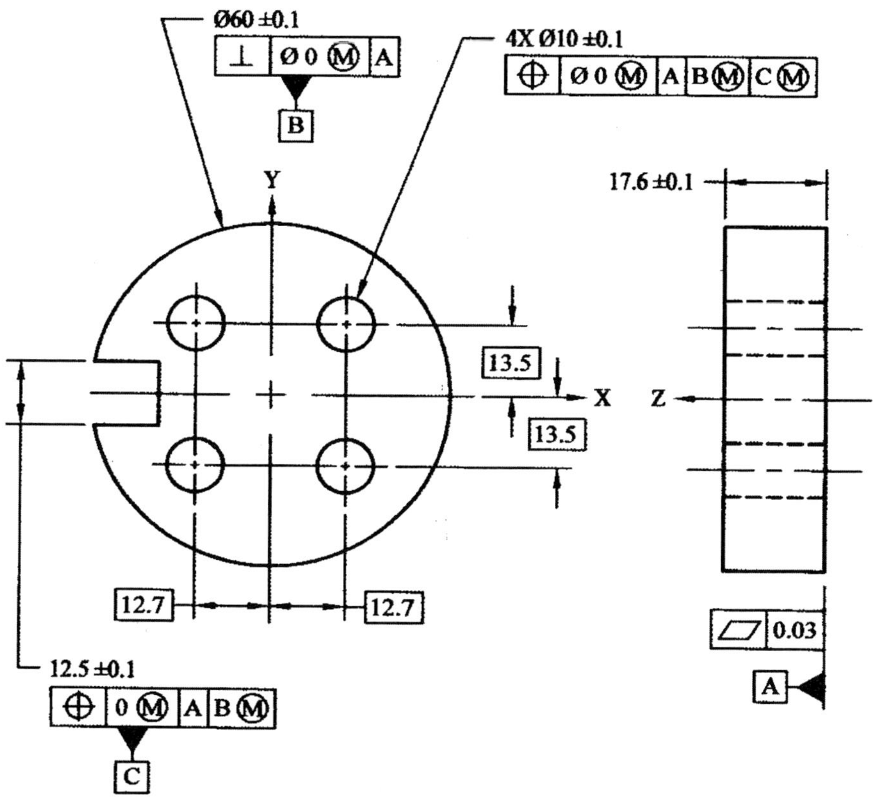

There are 6 spatial degrees of freedom. 3 are linear (x, y and z) and 3 are rotational (u, v and w). Which datum features control each of these? Specify the datum feature that controls each degree of spatial freedom listed below.

a. Movement in the x direction is controlled by datum **B.**

b. Movement in the y direction is controlled by datum **B.**

c. Movement in the z direction is controlled by datum **A.**

d. Rotation in the u direction is controlled by datum **A.**

e. Rotation in the v direction is controlled by datum **A.**

f. Rotation in the w direction is controlled by datum **C.**

Answers to workbook page 607

26.Overriding the Degrees of Spatial Freedom Controlled by Datums.

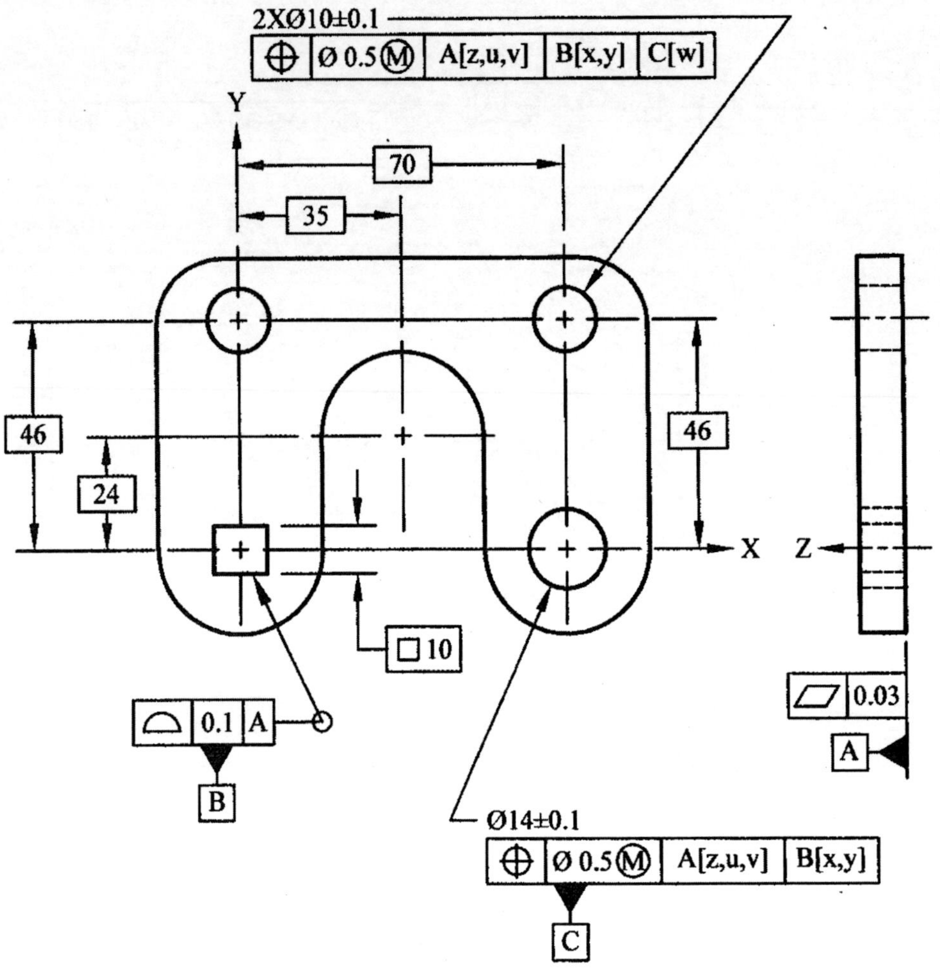

The positional controls shown in the illustration above designate which spatial degrees of freedom are controlled by the datum references. In the 4-hole positional control, what degrees of spatial freedom would each datum reference have controlled if not overridden by the bracketed designations?

a. Datum A would have controlled the following degrees of spatial freedom: **z, u, v**
b. Datum B would have controlled the following degrees of spatial freedom: **x, y, w**
c. Datum C would have controlled the following degrees of spatial freedom: **None**

Answers to workbook page 608

27. Please fill in the correct MMB or LMB in the blank spaces on the following illustrations.

Correct Maximum Material Boundary Calculations

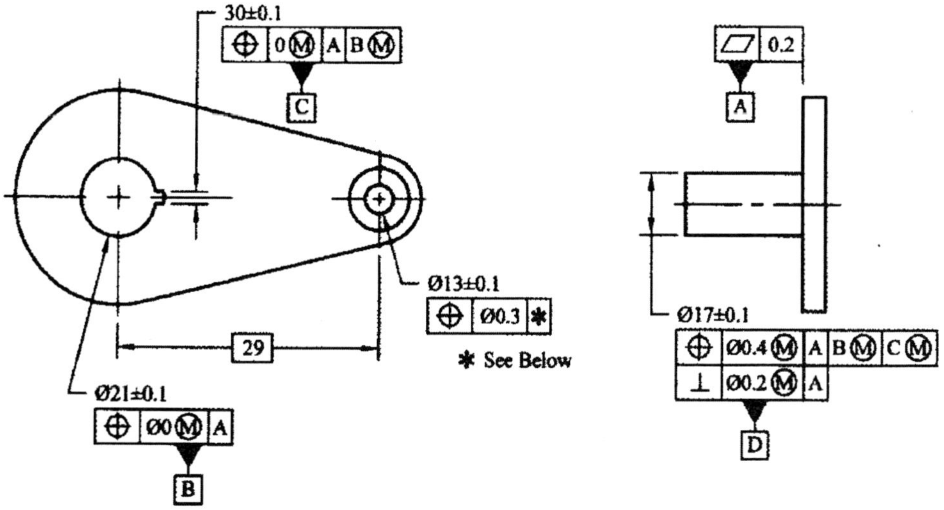

Option	Geometric Control	Maximum Material Boundary of datum feature D
(a)	⊕ Ø 0.3 DⓂ	17.1
(b)	⊕ Ø 0.3 A DⓂ	17.1+0.2 = 17.3
(c)	⊕ Ø 0.3 A B DⓂ	17.1+0.4 = 17.5

Correct Least Material Boundary Calculations

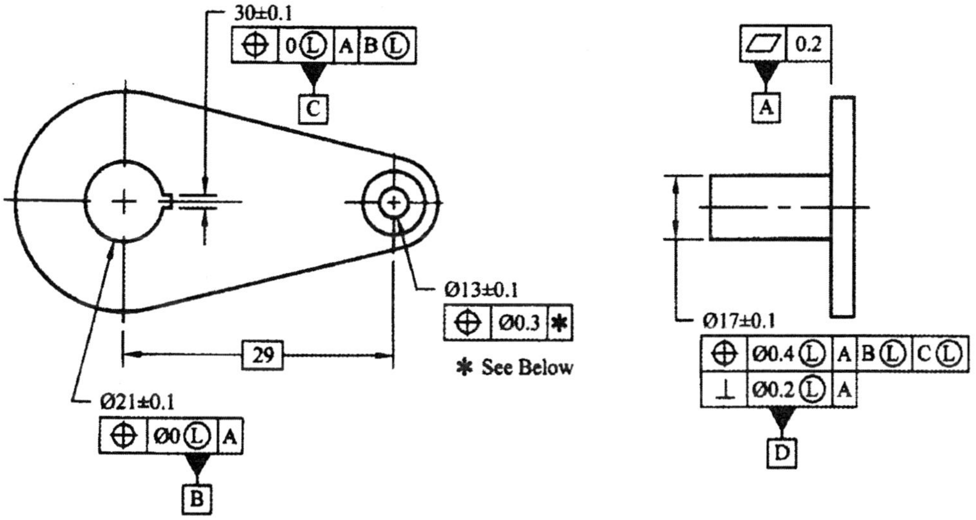

Option	Geometric Control	Least Material Boundary of datum feature D
(a)	⊕ Ø 0.3 DⓁ	16.9
(b)	⊕ Ø 0.3 A DⓁ	16.9-0.2 = 16.7
(c)	⊕ Ø 0.3 A B DⓁ	16.9-0.4 = 16.5

28. Answers to workbook page 609

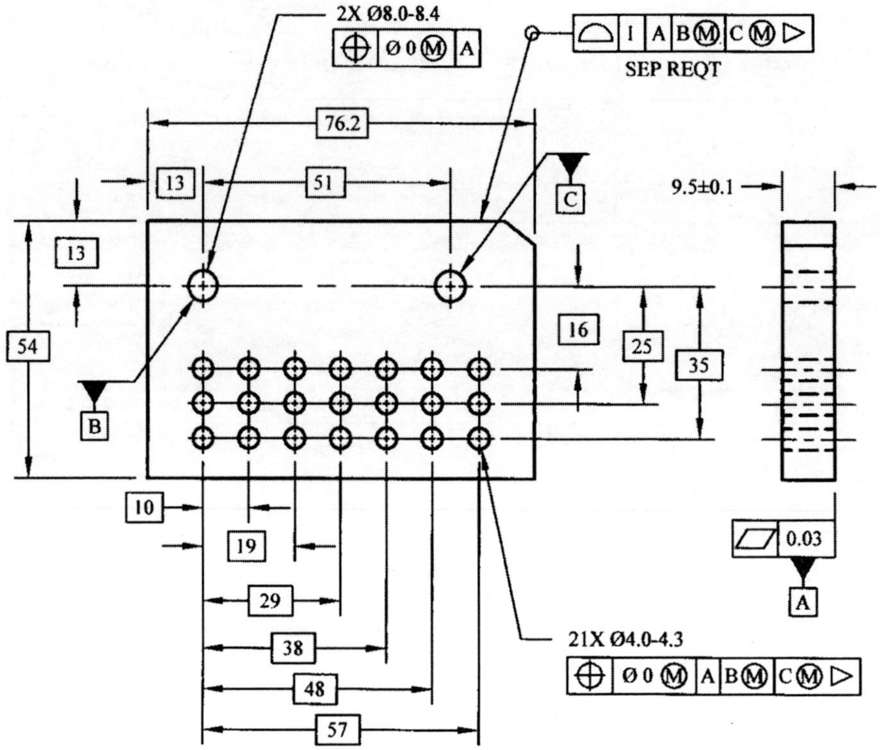

For the 21-hole pattern:

a. What does the symbol Ⓜ used after the zero tolerance in the feature control frame (FCF) stand for? **Maximum Material Condition**

b. Would the datum feature simulator for datum feature B be stationary or moveable? **Stationary**

c. What does the symbol Ⓜ used after datum feature B in the FCF stand for? **Maximum Material Boundary**

d. What does the symbol used after the Ⓜ after datum feature C stand for? **Translation**

e. What size gage pin would be used to represent datum feature B? **A dia. of 8**

f.. What size gage pin would be used to represent datum feature C? **A dia. of 8**

g. How would the datum feature simulator used to represent datum feature C differ from the one used for datum feature B? **The datum feature simulator for C would translate (be moveable) toward or away from the stationary datum feature B simulator.**

For the profile of a surface control:

h. What does the SEP REQT note stand for? **SEPARATE REQUIREMENT**

i. How would datum feature shift due to datum feature B's departure from its virtual condition differ if the SEP REQT note was not on the drawing? **With SEP REQT, the 21 holes (as a pattern) could shift in a different direction than the profile. Without SEP REQT, the 21 holes and the profile would constitute a SIMULTANEOUS REQUIREMENT and have to shift together.**

Answers to workbook page 610

29. Using the correct datum feature designations, feature control frames and local notes:
    a. Designate each of the 12 millimeter holes as a datum feature using one datum feature symbol (the letter D).
    b. Locate the smaller 6 millimeter holes to the left and right of each large hole to that 12 millimeter hole only, using datum plane A for perpendicularity, datum axis D for location and datum plane B for angular orientation.
    c. Use the local note 4X INDIVIUALLY twice.

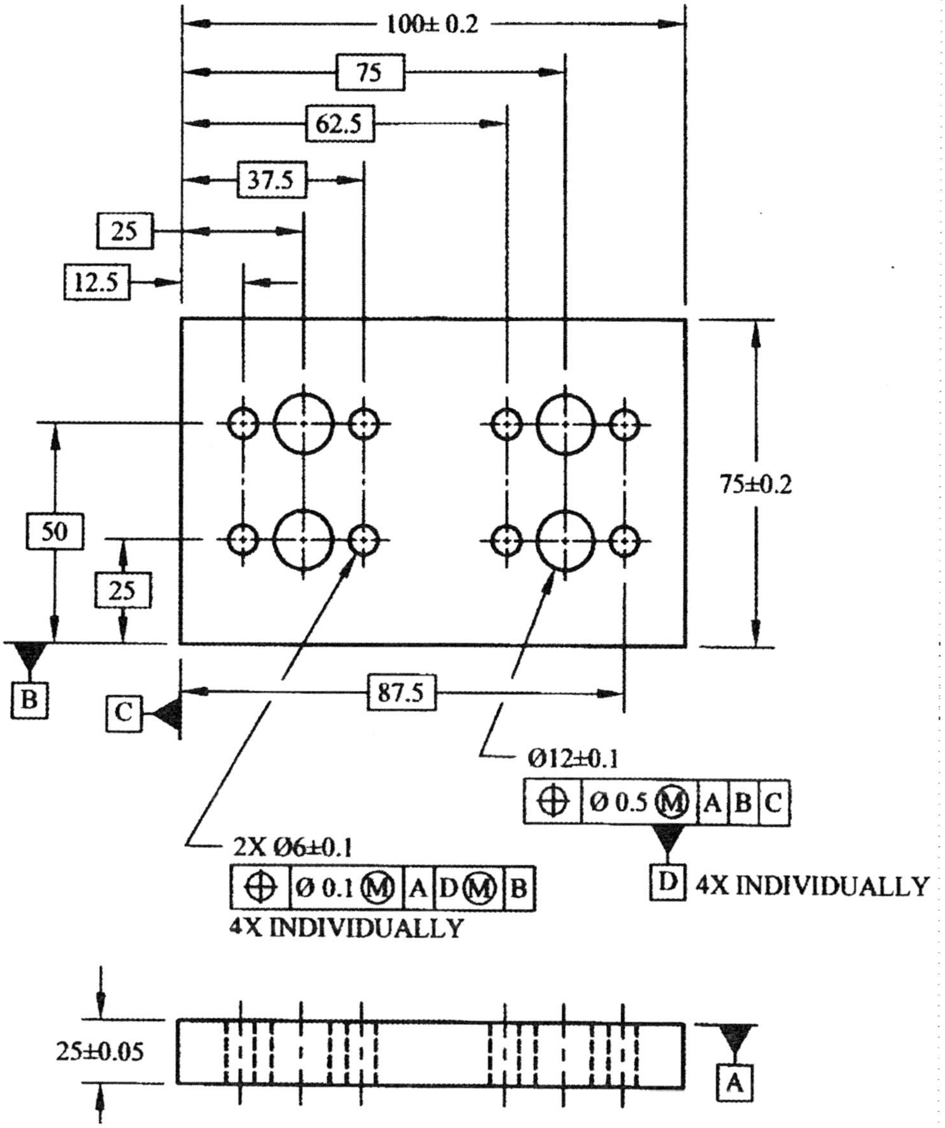

30. Answers to workbook page 611

Please apply a three-level composite positional tolerancing control to the three-hole pattern to mean the following:

1. The top level control locates the three holes to datums B and C. Regardless of the material boundary of datum feature B, the three holes must be centered to datum centerplane B and hold the 19 millimeter basic dimension from datum plane C to within a diameter of 3 millimeters if the holes are produced at their maximum material condition.
2. The middle level control is ultimately only holding the orientation of the three holes (as a pattern) to datum centerplane B and datum plane C to within a diameter of 1.5 millimeters if the holes are produced at their maximum material condition.
3. The third level control is aligning the three holes' axes to one another for coaxiality to within a diameter of 0.3 millimeters if the holes are produced at their maximum material condition. It is worth noting that no datum feature references are used here, in that, none are necessary to position the three holes to each other. This tolerance is the one that allows the 8 millimeter screw to pass through the three holes. It generates a virtual condition (inner boundary) for the three holes that is calculated by subtracting the 0.3 millimeter diameter position tolerance from the 8.3 millimeter maximum material condition of the holes. 8.3 minus 0.3 equals an 8 millimeter diameter virtual condition. This is the worst mating condition and is compatible with the 8 millimeter screw that must pass through. Since 8 millimeter screws are actually a little smaller than 8 millimeters, there is even a little clearance built into this assembly condition.

Detail Drawing

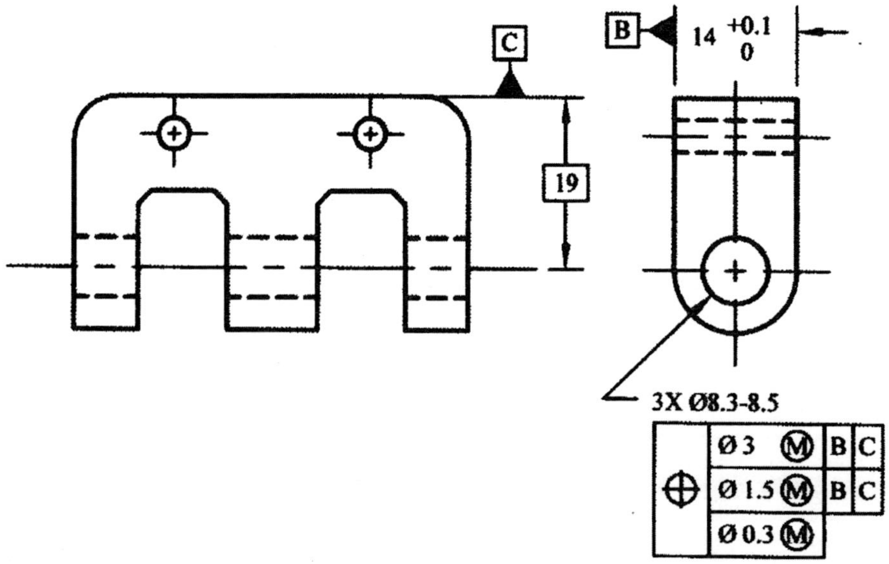

SECTION 8-Answers to pages 612-621

Workbook Answer to page 612

a) None, except position and a perpendicularity refinement might be easier to interpret. There is only one hole. No need for composite position.

b) Total wide tolerance zone.

c) Angular orientation – it stops pattern rotation.

d) Pattern could rotate with the only control being the general tolerance of ±5°.

Workbook Answer to page 613

Assembly

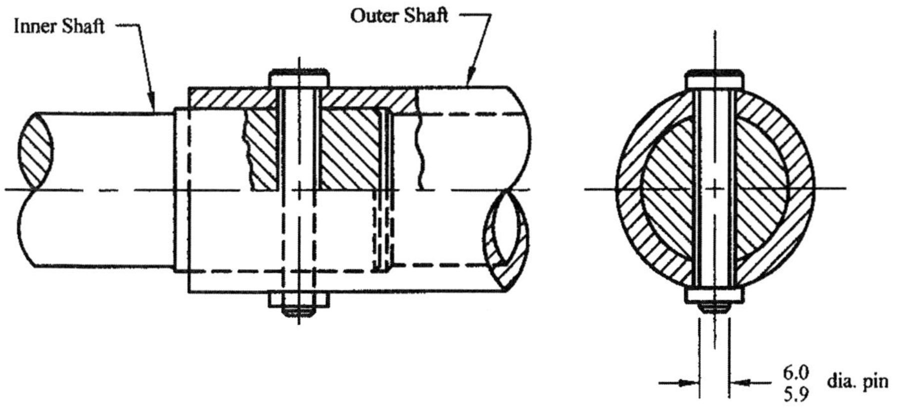

Detail drawing callouts

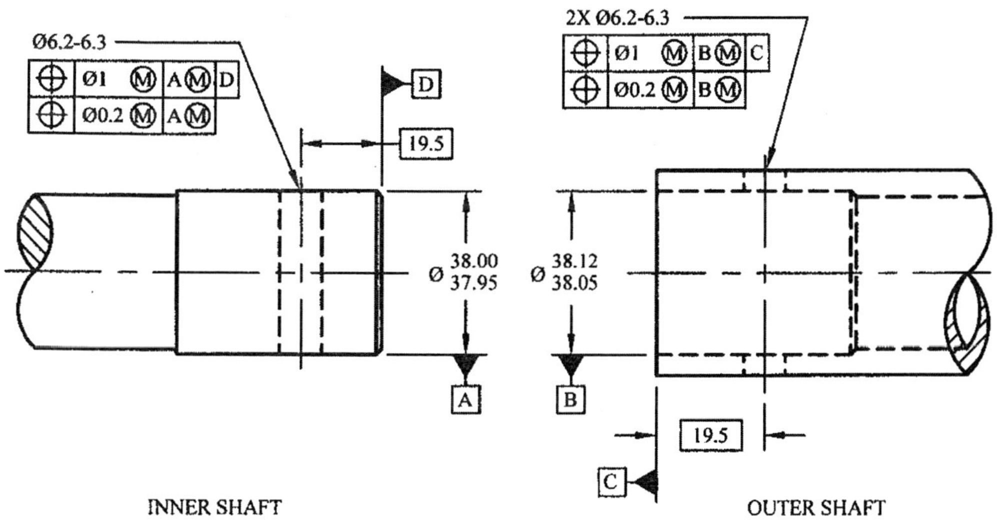

∠ = ±1°

Workbook Answer to pages 614 and 615

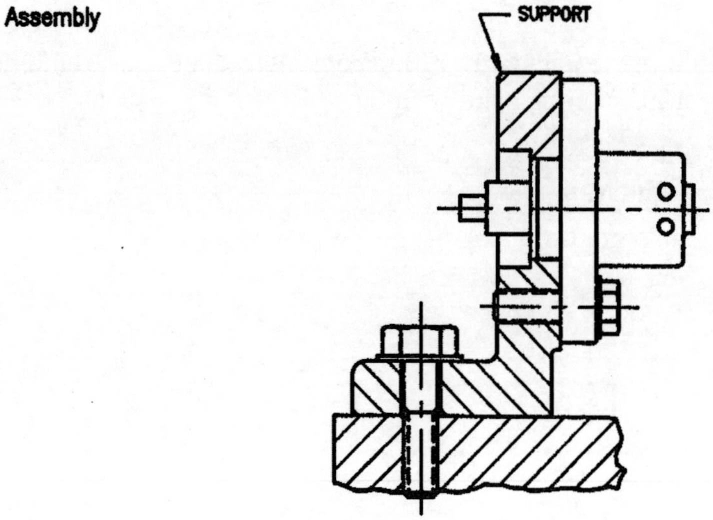

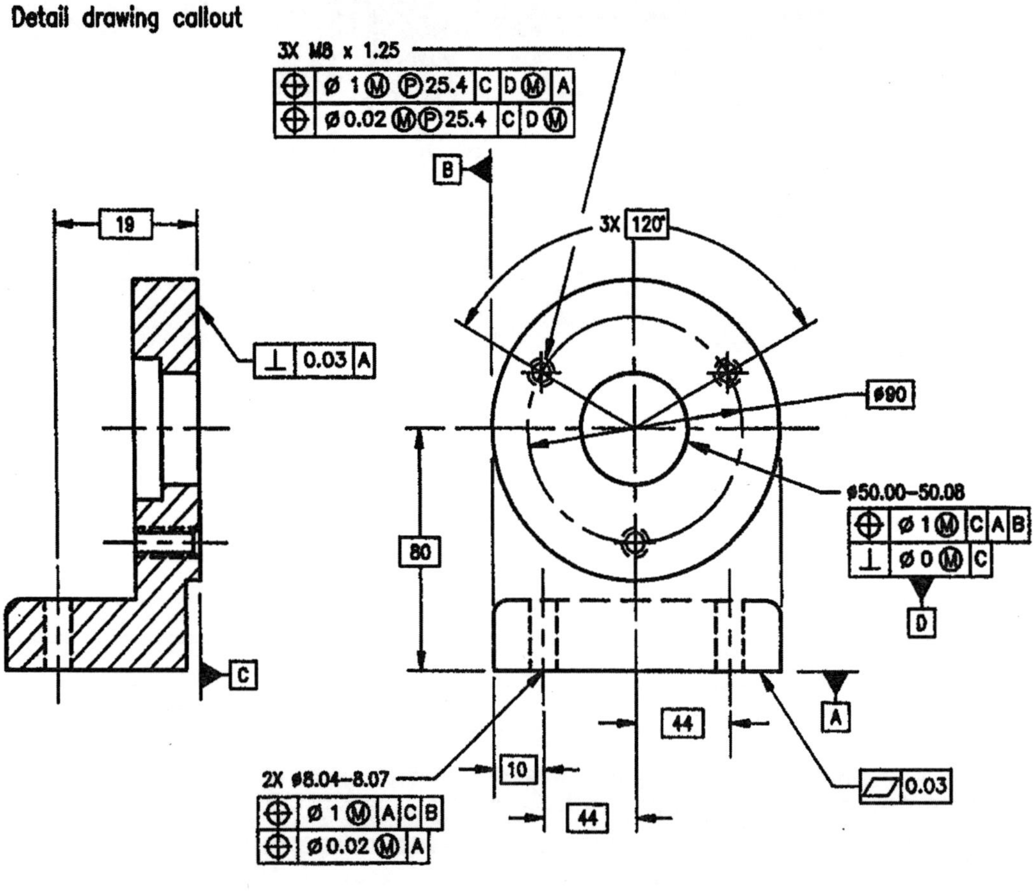

Workbook Answer to pages 616 and 617

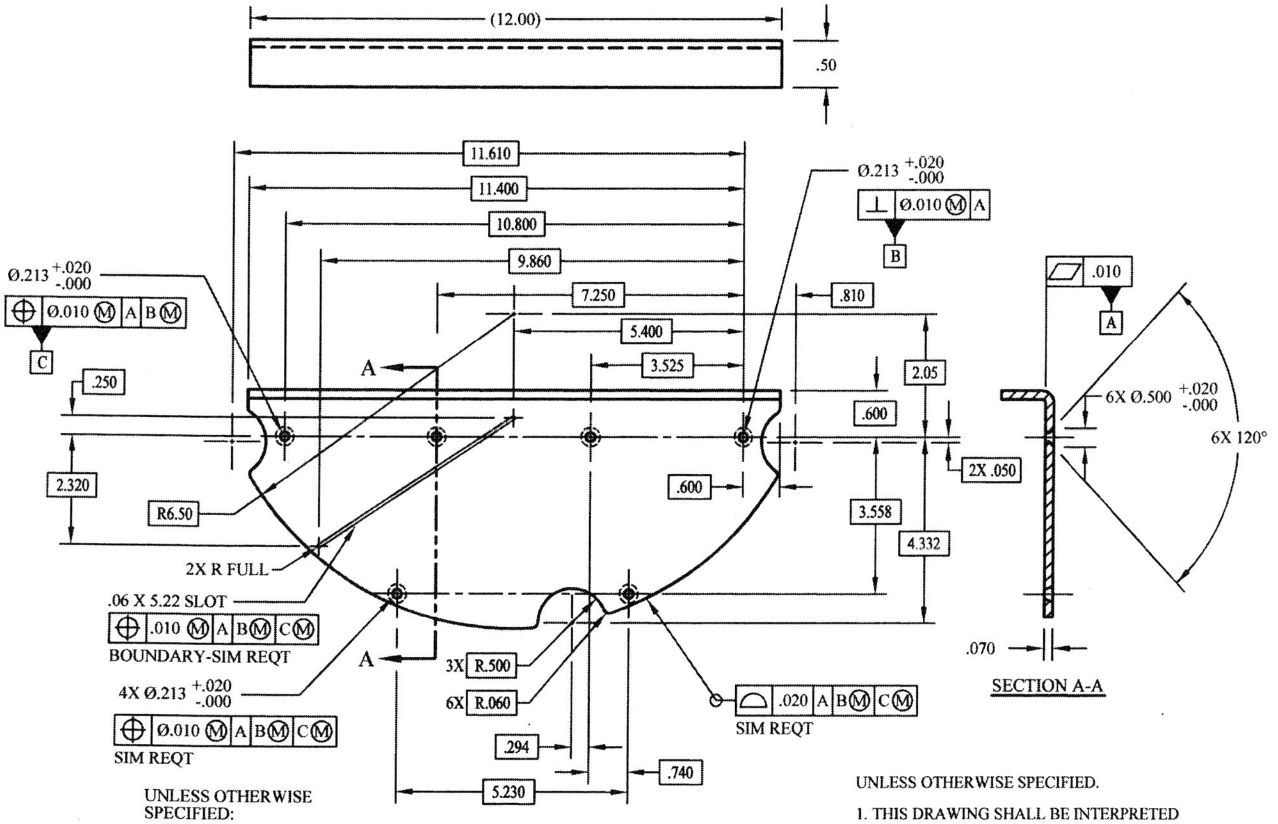

UNLESS OTHERWISE SPECIFIED:

.XX = ±.02
.XXX = ±.010
∠ = ±1°

UNLESS OTHERWISE SPECIFIED.

1. THIS DRAWING SHALL BE INTERPRETED PER ASME Y14.5M-1994.
2. BEND RADII TO BE MINIMUM.
3. EDGES TO BE BROKEN AND DEBURRED.

Workbook Answer to page 618

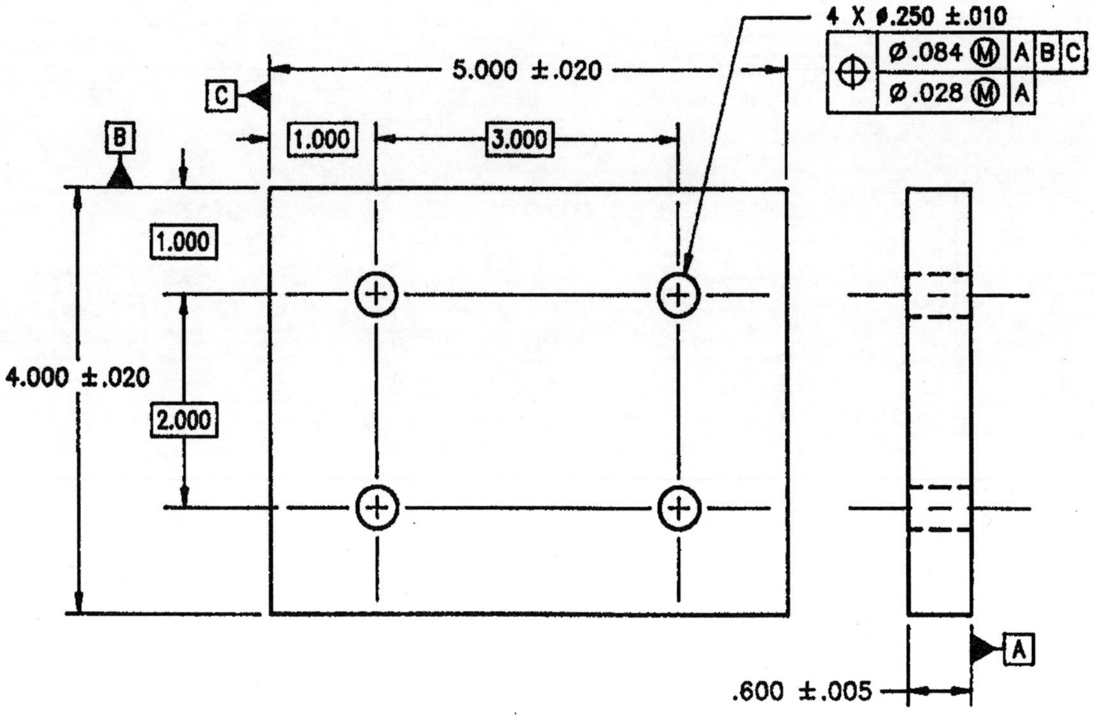

Workbook Answer to page 619

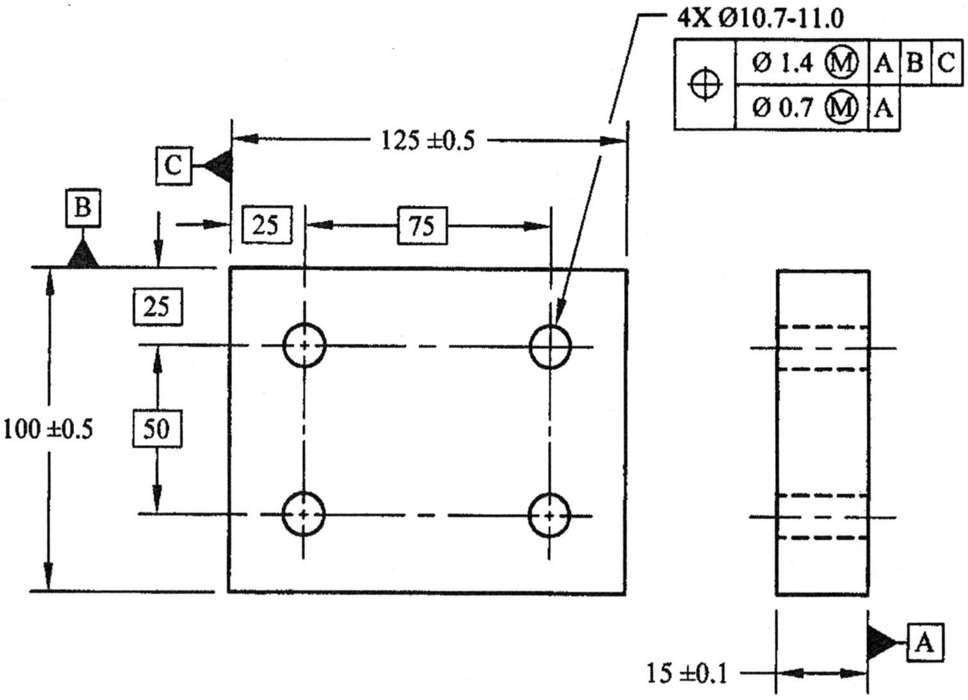

UNLESS OTHERWISE SPECIFIED
ALL ANGLES ARE ±1°

Workbook Answer to pages 620 and 621

Assembly

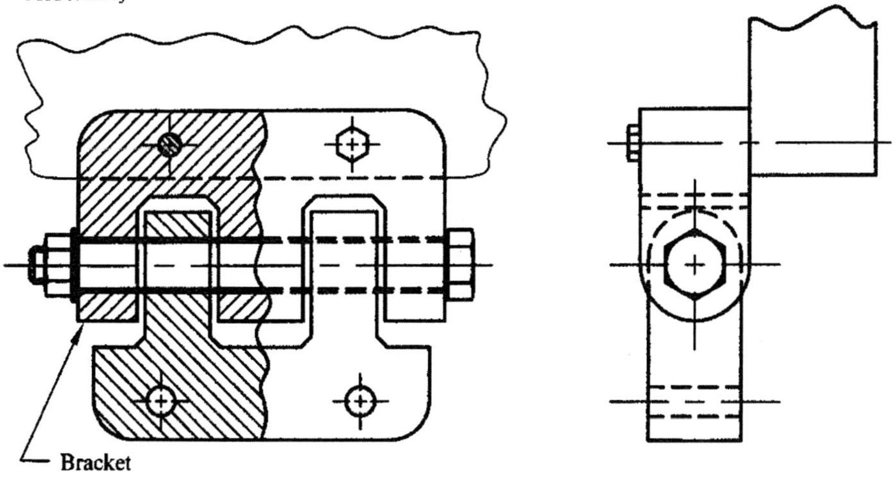

Detail Drawing Cutout

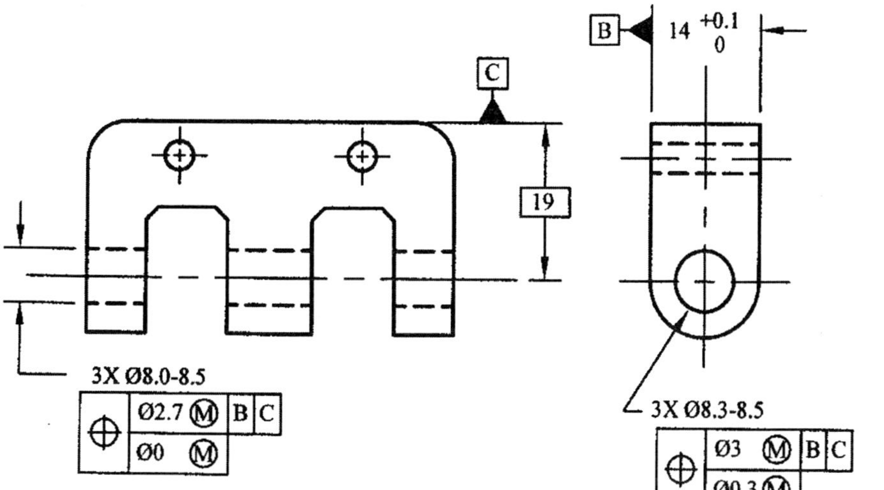

(Alternate design to answer question #2)

Answer to question #1

8.5=LMC of Holes
Step 7 + *3.0* =**Unknown Position Tol. at MMC**
+ 0.2=Bonus Tolerance
11.7=**Outer boundary of holes** *Step 6*

14.1=MMC of Datum Feature B
- *11.7*=**Outer Boundary of Holes** *Step 5*
2.4 =**Material on Both Sides of Holes** *Step 4*

2.4=**Material on both sides of hole divided by 2=Wall Thickness without pits=***1.2 Step 3*

1.2 =**Wall Thickness without pits** *Step 2*
- *0.1* =**Size/Flatness tolerance on datum feature B (Pits)** *Step 1*
1.1 =Minimum Wall Thickness

Numbers not shown in bold type and in italics are calculated from bold numbers that are given.

MMC Holes (8.3) minus MMC Screw (8.0) = *0.3* **Geo. Tol. to align the 3 holes**